Readings from
**SCIENTIFIC
AMERICAN**

PROGRESS IN
NEUROSCIENCE

With Introductions by
Richard F. Thompson
Stanford University

W. H. Freeman and Company
New York

Library of Congress Cataloging-in-Publication Data

Main entry under title:

Progress in neuroscience.

"Readings from Scientific American."
Bibliography: p.
Includes index.
1. Brain—Addresses, essays, lectures. 2. Neurophysiology—Addresses, essays, lectures. I. Thompson,
Richard F. II. Scientific American. [DNLM: 1. Neurology
—collected works. WL 5 P964]

QP376.P75 1986 612'.82 85-16120
ISBN 0-7167-1726-3
ISBN 0-7167-1727-1 (pbk.)

Printed in the United States of America

1234567890 KP 4321089876

CONTENTS

PREFACE

The human brain is the most complex structure in the known universe. The extraordinary properties of this three or so pounds of soft tissue have made it possible for *Homo sapiens* to dominate the earth, change the course of evolution through genetic engineering, walk on the moon, and create art and music of surpassing beauty. Furthermore, the limits of the human mind are not known. The challenge and excitement felt by neuroscientists, those who study the brain, are probably greater than in any other field of science. Many believe that the answers to the age-old questions of consciousness and knowledge will one day be found here.

Neuroscientists study all the aspects of the brain, its structure and development, the chemical and electrical phenomena occurring in its neurons and how these interact, and the brain's unique output, behavior and experience. The nervous system, particularly its anatomy and elementary functions, has been studied for centuries, but the field of neuroscience as a unified discipline is only a few years old.

Despite the brain's complexities, the basic principles governing the functioning of the nerve cells are not really very difficult to grasp, at least at our present level of knowledge. By the same token, the basic organization of the nerve cells into structures and systems can be understood at a relatively simple level, although the brain's detailed wiring is for the most part unknown.

These articles, selected from recent issues of *Scientific American*, present an overview of our understanding of the major fields of neuroscience and highlight exciting new discoveries and theories about the brain. They were also chosen because they complement the editor's book: *The Brain, an Introduction to Neuroscience*, New York, W. H. Freeman and Company, 1985. Though each volume can be read separately, when read together they form a cohesive text on the nervous system and on biologic psychology.

This volume is divided into five sections. The first, *Neuron and Synapse*, focusses on the brain's basic functional element, the neuron. The neuron's most important function, communicating with other neurons or target cells by the chemical transmission of information at synapses, is addressed in the second section, *Chemistry of the Brain*. Study of the brain's chemistry is the newest and most rapidly advancing field in neuroscience.

How the complex structure of the brain grows and develops from a single cell is a question that has attracted much interest, and it forms the third section: *Development of the Brain. Sensory Processes*, the fourth section, examines the visual and auditory systems, the major gateways that provide the brain with information about the world. Our understanding about the brain substrates of vision and visual sensation is more advanced than that of all other brain functions. Finally, in the fifth section, *Brain, Behavior, and the Mind,* three topics are chosen from the vast field of brain function and behavior—a newly discovered instance of how genes can control a simple behavior in an organism with a simple nervous system, the etiology and treatment of mania, and perhaps the ultimate question, the mind-body problem.

This book of readings contains only up-to-date information on current topics. The articles, prepared by outstanding scientists in several fields of neuroscience, are clearly written and understandable to an interested reader who may have no special background or training in the field. Characteristic of *Scientific American* is the unique way in which work on the forefronts of scientific knowledge is presented with accuracy and, at the same time, in a highly readable and interesting manner.

A brief bibliography is provided at the end of the book to help those desiring to pursue a topic further. Cross-references within the articles are consistent with the following conventions: A reference to an article included in this book is noted by the title of the article and the page on which it begins; a reference to an article that is available as a *Scientific American* offprint, but is not included here, is noted by the article's title and offprint number; and a reference to an article published by *Scientific American*, but not available as an offprint, is noted by the title of the article and the month and year of its publication.

Richard F. Thompson

I

NEURON AND SYNAPSE

I   NEURON AND SYNAPSE

INTRODUCTION

The functional unit of the brain is the nerve cell or neuron. The human brain is thought to consist of perhaps 100 billion (100,000,000,000, or 10^{11}) individual neurons, each a separate cell. This fact, now taken for granted, only became firmly established at the beginning of the 20th century. Before then, many anatomists believed the brain was an exception to the basic biologic principle that all tissue is made up of individual cells.

If a piece of brain tissue is stained with a substance that colors all the parts of its cells, it looks like a continuous mass, a tangled web of fibers with cell nuclei scattered throughout. In the late 19th century, the anatomist Camillo Golgi discovered a stain that colored only an occasional neuron in brain tissue but colored it completely. This stain made it possible to see complete neurons, with all their processes. The discovery of the Golgi stain came about, according to the story, when a cleaning woman disposed of a piece of brain tissue from Golgi's desk in a waste bucket that contained silver nitrate solution. When Golgi returned and found the tissue, it had undergone the first successful Golgi stain.

Interestingly, Golgi himself did not believe in the "neuron doctrine," that that brain was composed of individual neurons. Another anatomist, Rámon y Cajal, systematically applied the Golgi stain to animal brains and established that all the parts of the brain are composed of individual neurons. Cajal began the immensely complex task of diagramming the brain's wiring, the patterns of interconnections among the neurons.

A typical neuron consists of a cell body containing the nucleus and a number of fibers extending from it. The neuron transmits information to other cells by sending activity out just one fiber, the axon. All the other fibrous extensions of the cell body, the dendrites, receive information from other neurons. The axons of some neurons in the human body are more than a meter in length; other axons are not much longer than dendrites, which are tens to hundreds of micrometers in length.

The book begins with an in-depth examination of the neuron by Charles F. Stevens of Yale University School of Medicine. Dr. Stevens focuses on the nerve cell membrane and how it became specialized at transmitting information out its axon and triggering the release of chemical transmitter substance at its synapses. The synapse is the functional connection between the axon terminal and another neuron; it is the point where information is transmitted from one neuron to another. A very tiny space, called the synaptic cleft, separates the axon terminal and the cell body, or dendrite of the cell, with which it synapses.

A given neuron in the brain may receive several thousand synaptic connections from other neurons. Hence, if the human brain has 10^{11} neurons, it has at least 10^{14} synapses, or many trillions. The number of *possible* different combinations of synaptic connections in a single human brain is greater than the total number of atomic particles in the known universe. The diversity of possible interconnections in a human brain, therefore, are apparently without limit.

The basic process of chemical synaptic transmission is straight-forward. An action potential is conducted out the axon to the synaptic terminal where it triggers the release of the transmitter chemical from the terminal. It does this by triggering an increase in the conductance of calcium ions into the terminal. The process of synaptic transmission and the key role of calcium is examined in the article by Rodolfo Llinás of New York University Medical School.

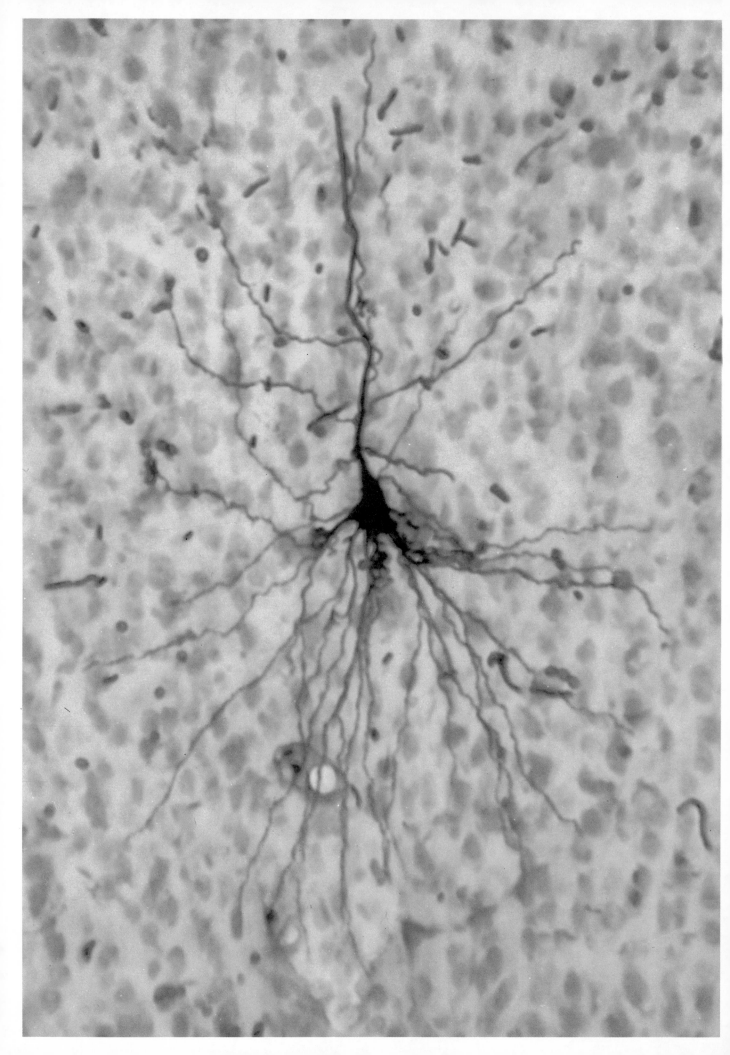

The Neuron

1

by Charles F. Stevens
September 1979

It is the individual nerve cell, the building block of the brain.
It transmits nerve impulses over a single long fiber (the axon)
and receives them over numerous short fibers (the dendrites)

Neurons, or nerve cells, are the building blocks of the brain. Although they have the same genes, the same general organization and the same biochemical apparatus as other cells, they also have unique features that make the brain function in a very different way from, say, the liver. The important specializations of the neuron include a distinctive cell shape, an outer membrane capable of generating nerve impulses, and a unique structure, the synapse, for transferring information from one neuron to the next.

The human brain is thought to consist of 10^{11} neurons, about the same number as the stars in our galaxy. No two neurons are identical in form. Nevertheless, their forms generally fall into only a few broad categories, and most neurons share certain structural features that make it possible to distinguish three regions of the cell: the cell body, the dendrites and the axon. The cell body contains the nucleus of the neuron and the biochemical machinery for synthesizing enzymes and other molecules essential to the life of the cell. Usually the cell body is roughly spherical or pyramid-shaped. The dendrites are delicate tube-like extensions that tend to branch repeatedly and form a bushy tree around the cell body. They provide the main physical surface on which the neuron receives incoming signals. The axon extends away from the cell body and provides the pathway over which signals can travel from the cell body for long distances to other parts of the brain and

the nervous system. The axon differs from the dendrites both in structure and in the properties of its outer membrane. Most axons are longer and thinner than dendrites and exhibit a different branching pattern: whereas the branches of dendrites tend to cluster near the cell body, the branches of axons tend to arise at the end of the fiber where the axon communicates with other neurons.

The functioning of the brain depends on the flow of information through elaborate circuits consisting of networks of neurons. Information is transferred from one cell to another at specialized points of contact: the synapses. A typical neuron may have anywhere from 1,000 to 10,000 synapses and may receive information from something like 1,000 other neurons. Although synapses are most often made between the axon of one cell and the dendrite of another, there are other kinds of synaptic junction: between axon and axon, between dendrite and dendrite and between axon and cell body.

At a synapse the axon usually enlarges to form a terminal button, which is the information-delivering part of the junction. The terminal button contains tiny spherical structures called synaptic vesicles, each of which can hold several thousand molecules of chemical transmitter. On the arrival of a nerve impulse at the terminal button, some of the vesicles discharge their contents into the narrow cleft that separates the button from the membrane of another cell's dendrite, which is designed to receive

the chemical message. Hence information is relayed from one neuron to another by means of a transmitter. The "firing" of a neuron—the generation of nerve impulses—reflects the activation of hundreds of synapses by impinging neurons. Some synapses are excitatory in that they tend to promote firing, whereas others are inhibitory and so are capable of canceling signals that otherwise would excite a neuron to fire.

Although neurons are the building blocks of the brain, they are not the only kind of cell in it. For example, oxygen and nutrients are supplied by a dense network of blood vessels. There is also a need for connective tissue, particularly at the surface of the brain. A major class of cells in the central nervous system is the glial cells, or glia. The glia occupy essentially all the space in the nervous system not taken up by the neurons themselves. Although the function of the glia is not fully understood, they provide structural and metabolic support for the delicate meshwork of the neurons.

One other kind of cell, the Schwann cell, is ubiquitous in the nervous system. All axons appear to be jacketed by Schwann cells. In some cases the Schwann cells simply enclose the axon in a thin layer. In many cases, however, the Schwann cell wraps itself around the axon in the course of embryonic development, giving rise to the multiple dense layers of insulation known as myelin. The myelin sheath is interrupted every millimeter or so along the axon by narrow gaps called the nodes of Ranvier. In axons that are sheathed in this way the nerve impulse travels by jumping from node to node, where the extracellular fluid can make direct contact with the cell membrane. The myelin sheath seems to have evolved as a means of conserving the neuron's metabolic energy. In general myelinated nerve fibers conduct nerve impulses faster than unmyelinated fibers.

Neurons can work as they do because their outer membranes have special

NEURON FROM A CAT'S VISUAL CORTEX has been labeled in the photomicrograph on the opposite page by injection with the enzyme horseradish peroxidase. The cell bodies in the background are counterstained with a magenta dye. All the fibers extending from the cell body are dendrites, which receive information from other neurons. The fiber that transmits information, the axon, is much finer and not readily visible at this magnification. The thickest fiber, extending vertically upward, is known as the apical dendrite, only a small portion of which falls within this section. At this magnification (about 500 diameters) the complete apical dendrite would be about 75 centimeters long. (It can be traced through adjacent sections.) The activity of this particular cell was recorded in the living animal and was found to respond optimally to a light-dark border rotated about 60 degrees from the vertical. The neuron is classified as a pyramidal cell because of its form. It is one of two major types in cortex of mammals. Micrograph was made by Charles Gilbert and Torsten N. Wiesel of Harvard Medical School.

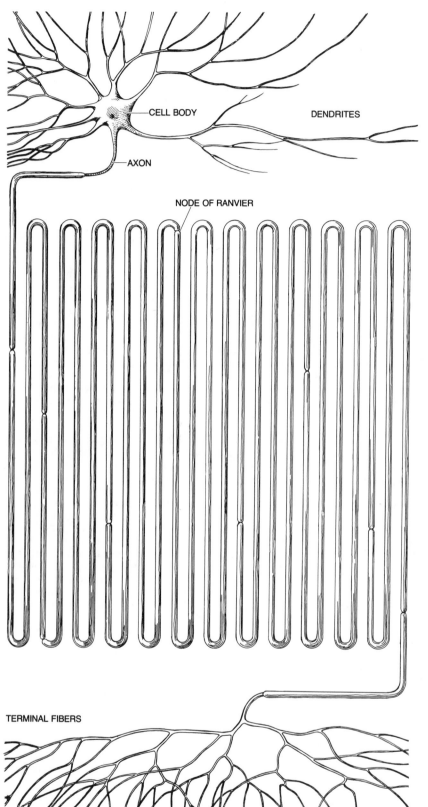

CELL BODY

DENDRITES

AXON

NODE OF RANVIER

TERMINAL FIBERS

TYPICAL NEURON of a vertebrate animal can carry nerve impulses for a considerable distance. The neuron depicted here, with its various parts drawn to scale, is enlarged 250 times. The nerve impulses originate in the cell body and are propagated along the axon, which may have one or more branches. This axon, which is folded for diagrammatic purposes, would be a centimeter long at actual size. Some axons are more than a meter long. The axon's terminal branches form synapses with as many as 1,000 other neurons. Most synapses join the axon terminals of one neuron with the dendrites forming a "tree" around the cell body of another neuron. Thus the dendrites surrounding the neuron in the diagram might receive incoming signals from tens, hundreds or even thousands of other neurons. Many axons, such as this one, are insulated by a myelin sheath interrupted at intervals by the regions known as nodes of Ranvier.

properties. Along the axon the membrane is specialized to propagate an electrical impulse. At the terminal of the axon the membrane releases transmitters, and on the dendrites it reponds to transmitters. In addition the membrane mediates the recognition of other cells in embryonic development, so that each cell finds its proper place in the network of 10^{11} cells. Much recent investigation therefore focuses on the membrane properties responsible for the nerve impulse, for synaptic transmission, for cell-cell recognition and for structural contacts between cells.

The neuron membrane, like the outer membrane of all cells, is about five nanometers thick and consists of two layers of lipid molecules arranged with their hydrophilic ends pointing toward the water on the inside and outside of the cell and with their hydrophobic ends pointing away from the water to form the interior of the membrane. The lipid parts of the membrane are about the same for all kinds of cells. What makes one cell membrane different from another are various specific proteins that are associated with the membrane in one way or another. Proteins that are actually embedded in the lipid bilayer are termed intrinsic proteins. Other proteins, the peripheral membrane proteins, are attached to the membrane surface but do not form an integral part of its structure. Because the membrane lipid is fluid even the intrinsic proteins are often free to move by diffusion from place to place. In some instances, however, the proteins are firmly fastened down by a substructure.

The membrane proteins of all cells fall into five classes: pumps, channels, receptors, enzymes and structural proteins. Pumps expend metabolic energy to move ions and other molecules against concentration gradients in order to maintain appropriate concentrations of these molecules within the cell. Because charged molecules do not pass through the lipid bilayer itself cells have evolved channel proteins that provide selective pathways through which specific ions can diffuse. Cell membranes must recognize and attach many types of molecules. Receptor proteins fulfill these functions by providing binding sites with great specificity and high affinity. Enzymes are placed in or on the membrane to facilitate chemical reactions at the membrane surface. Finally, structural proteins both interconnect cells to form organs and help to maintain subcellular structure. These five classes of membrane proteins are not necessarily mutually exclusive. For example, a particular protein might simultaneously be a receptor, an enzyme and a pump.

Membrane proteins are the key to understanding neuron function and therefore brain function. Because they play such a central role in modern views of the neuron, I shall organize my discussion around a description of an ion pump, various types of channel and some other proteins that taken together endow neurons with their unique properties. The general idea will be to summarize the important characteristics of the membrane proteins and to explain how these characteristics account for the nerve impulse and other complex features of neuron function.

Like all cells the neuron is able to maintain within itself a fluid whose composition differs markedly from that of the fluid outside it. The difference is particularly striking with regard to the concentration of the ions of sodium and potassium. The external medium is about 10 times richer in sodium than the internal one, and the internal medium is about 10 times richer in potassium than the external one. Both sodium and potassium leak through pores in the cell membrane, so that a pump must operate continuously to exchange sodium ions that have entered the cell for potassium ions outside it. The pumping is accomplished by an intrinsic membrane protein called the sodium-potassium adenosine triphosphatase pump, or more often simply the sodium pump.

The protein molecule (or complex of protein subunits) of the sodium pump has a molecular weight of about 275,000 daltons and measures roughly six by eight nanometers, or slightly more than the thickness of the cell membrane. Each sodium pump can harness the energy stored in the phosphate bond of adenosine triphosphate (ATP) to exchange three sodium ions on the inside of the cell for two potassium ions on the outside. Operating at the maximum rate, each pump can transport across the membrane some 200 sodium ions and 130 potassium ions per second. The actual rate, however, is adjusted to meet the needs of the cell. Most neurons have between 100 and 200 sodium pumps per square micrometer of membrane surface, but in some parts of their surface the density is as much as 10 times higher. A typical small neuron has perhaps a million sodium pumps with a capacity to move about 200 million sodium ions per second. It is the transmembrane gradients of sodium and potassium ions that enable the neuron to propagate nerve impulses.

Membrane proteins that serve as channels are essential for many aspects of neuron function, particularly for the nerve impulse and synaptic transmission. As an introduction to the role played by channels in the electrical activity of the brain I shall briefly describe the mechanism of the nerve impulse and then return to a more systematic survey of channel properties.

Since the concentration of sodium and potassium ions on one side of the cell membrane differs from that on the

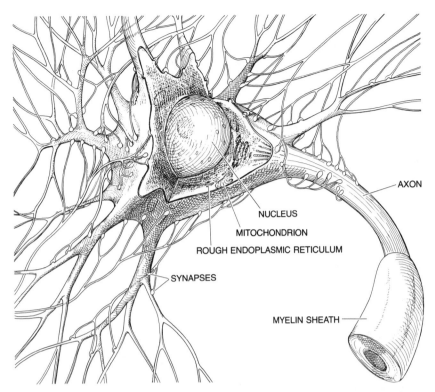

CELL BODY OF A NEURON incorporates the genetic material and complex metabolic apparatus common to all cells. Unlike most other cells, however, neurons do not divide after embryonic development; an organism's original supply must serve a lifetime. Projecting from the cell body are several dendrites and a single axon. The cell body and dendrites are covered by synapses, knoblike structures where information is received from other neurons. Mitochondria provide the cell with energy. Proteins are synthesized on the endoplasmic reticulum. A transport system moves proteins and other substances from cell body to sites where they are needed.

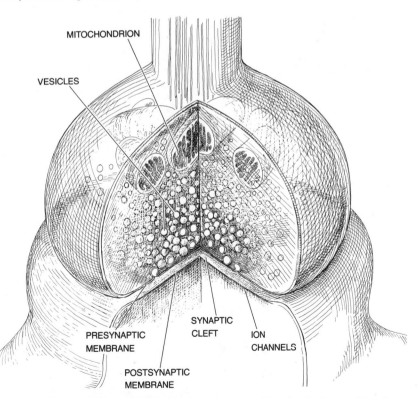

SYNAPSE is the relay point where information is conveyed by chemical transmitters from neuron to neuron. A synapse consists of two parts: the knoblike tip of an axon terminal and the receptor region on the surface of another neuron. The membranes are separated by a synaptic cleft some 200 nanometers across. Molecules of chemical transmitter, stored in vesicles in the axon terminal, are released into the cleft by arriving nerve impulses. Transmitter changes electrical state of the receiving neuron, making it either more likely or less likely to fire an impulse.

other side, the interior of the axon is about 70 millivolts negative with respect to the exterior. In their classic studies of nerve-impulse transmission in the giant axon of the squid a quarter of a century ago, A. L. Hodgkin, A. F. Huxley and Bernhard Katz of Britain demonstrated that the propagation of the nerve impulse coincides with sudden changes in the permeability of the axon membrane to sodium and potassium ions. When a nerve impulse starts at the origin of the axon, having been triggered in most cases by the cell body in response to dendritic synapses, the voltage difference across the axon membrane is locally lowered. Immediately ahead of the electrically altered region (in the direction in which the nerve impulse is propagated) channels in the membrane open and let sodium ions pour into the axon.

The process is self-reinforcing: the flow of sodium ions through the membrane opens more channels and makes it easier for other ions to follow. The sodium ions that enter change the internal potential of the membrane from negative to positive. Soon after the sodium channels open they close, and another group of channels open that let potassi-

um ions flow out. This outflow restores the voltage inside the axon to its resting value of −70 millivolts. The sharp positive and then negative charge, which shows up as a "spike" on an oscilloscope, is known as the action potential and is the electrical manifestation of the nerve impulse. The wave of voltage sweeps along until it reaches the end of the axon much as a flame travels along the fuse of a firecracker.

This brief description of the nerve impulse illustrates the importance of channels for the electrical activity of neurons and underscores two fundamental properties of channels: selectivity and gating. I shall discuss these two properties in turn. Channels are selectively permeable and selectivities vary widely. For example, one type of channel lets sodium ions pass through and largely excludes potassium ions, whereas another type of channel does the reverse. The selectivity, however, is seldom absolute. One type of channel that is fairly nonselective allows the passage of about 85 sodium ions for every 100 potassium ions; another more selective type passes only about seven sodium ions for every 100 potassium ions. The

first type, known as the acetylcholine-activated channel, has a pore about .8 nanometer in diameter that is filled with water. The second type, known as the potassium channel, has a much smaller opening and contains less water.

The sodium ion is about 30 percent smaller than the potassium ion. The exact molecular structure that enables the larger ion to pass through the cell membrane more readily than the smaller one is not known. The general principles that underlie the discrimination, however, are understood. They involve interactions between ions and parts of the channel structure in conjunction with a particular ordering of water molecules within the pore.

The gating mechanism that regulates the opening and closing of membrane channels takes two main forms. One type of channel, mentioned above in the description of the nerve impulse, opens and closes in response to voltage differences across the cell membrane; it is therefore said to be voltage-gated. A second type of channel is chemically gated. Such channels respond only slightly if at all to voltage changes but open when a particular molecule—a transmitter—binds to a receptor region

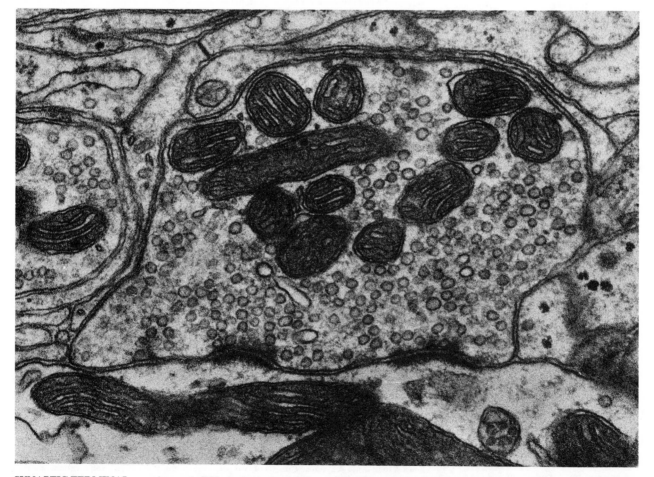

SYNAPTIC TERMINAL occupies most of this electron micrograph made by John E. Heuser of the University of California School of Medicine in San Francisco and Thomas S. Reese of the National Institutes of Health. The cleft separating the presynaptic membrane from the postsynaptic one undulates across the lower part of the picture. The large dark structures are mitochondria. The many round bodies are vesicles that hold transmitter. The fuzzy dark thickenings along the cleft are thought to be principal sites of transmitter release.

on the channel protein. Chemically gated channels are found in the receptive membranes of synapses and are responsible for translating the chemical signals produced by axon terminals into ion permeability changes during synaptic transmission. It is customary to name chemically gated channels according to their normal transmitter. Hence one speaks of acetylcholine-activated channels or GABA-activated channels. (GABA is gamma-aminobutyric acid.) Voltage-gated channels are generally named for the ion that passes through the channel most readily.

Proteins commonly change their shape as they function. Such alterations in shape, known as conformational changes, are dramatic for the contractile proteins responsible for cell motion, but they are no less important in many enzymes and other proteins. Conformational changes in channel proteins form the basis for gating as they serve to open and close the channel by slight movements of critically placed portions of the molecule that unblock and block the pore.

When either voltage-gated or chemically gated channels open and allow ions to pass, one can measure the resulting electric current. Quite recently it has become possible in a few instances to record the current flowing through a single channel, so that the opening and closing can be directly detected. One finds that the length of time a channel stays open varies randomly because the opening and closing of the channel represents a change in the conformation of the protein molecule embedded in the membrane. The random nature of the gating process arises from the haphazard collision of water molecules and other molecules with the structural elements of the channel.

In addition to ion pumps and channels neurons depend on other classes of membrane proteins for carrying out essential nervous-system functions. One of the important proteins is the enzyme adenylate cyclase, which helps to regulate the intracellular substance cyclic adenosine monophosphate (cyclic AMP). Cyclic nucleotides such as cyclic AMP take part in cell functions whose mechanisms are not yet understood in detail. The membrane enzyme adenylate cyclase appears to have two chief subunits, one catalytic and the other regulatory. The catalytic subunit promotes the formation of cyclic AMP. Various regulatory subunits, which are thought to be physically distinct from the catalytic one, can bind specific molecules (including transmitters that open and close channels) in order to control intracellular levels of cyclic AMP. The various types of regulatory subunit are named according to the molecule that normally binds to them; one, for example, is called serotonin-activated ade-

ACETYLCHOLINE-ACTIVATED CHANNELS are densely packed in the postsynaptic membrane of a cell in the electric organ of a torpedo, a fish that can administer an electric shock. This electron micrograph shows the platinum-plated replica of a membrane that had been frozen and etched. The size of the platinum particles limits the resolution to features larger than about two nanometers. According to recent evidence the channel protein molecule, which measures 8.5 nanometers across, consists of five subunits surrounding a channel whose narrowest dimension is .8 nanometer. The micrograph was made by Heuser and S. R. Salpeter.

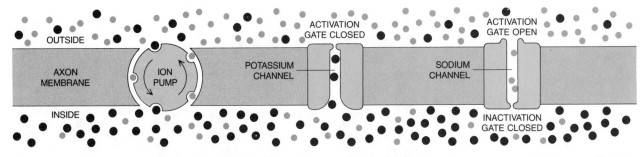

AXON MEMBRANE separates fluids that differ greatly in their content of sodium ions (*colored dots*) and potassium ions (*black dots*). The exterior fluid is about 10 times richer in sodium ions than in potassium ions; in the interior fluid the ratio is the reverse. The membrane is penetrated by proteins that act as selective channels for preferentially passing either sodium or potassium ions. In the resting state, when no nerve impulse is being transmitted, the two types of channel are closed and an ion pump maintains the ionic disequilibrium by pumping out sodium ions in exchange for potassium ions. The interior of the axon is normally about 70 millivolts negative with respect to the exterior. If this voltage difference is reduced by the arrival of a nerve impulse, the sodium channel opens, allowing sodium ions to flow into the axon. An instant later the sodium channel closes and the potassium channel opens, allowing an outflow of potassium ions. The sequential opening and closing of the two kinds of channel effects the propagation of the nerve impulse, which is illustrated below.

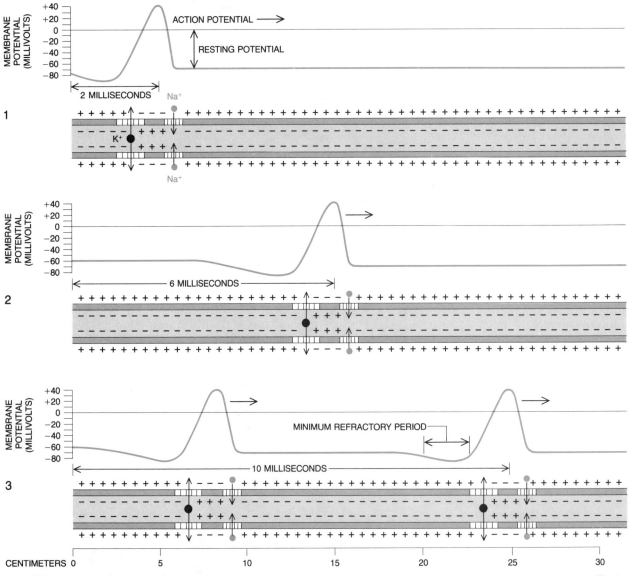

PROPAGATION OF NERVE IMPULSE along the axon coincides with a localized inflow of sodium ions (Na^+) followed by an outflow of potassium ions (K^+) through channels that are "gated," or controlled, by voltage changes across the axon membrane. The electrical event that sends a nerve impulse traveling down the axon normally originates in the cell body. The impulse begins with a slight depolarization, or reduction in the negative potential, across the membrane of the axon where it leaves the cell body. The slight voltage shift opens some of the sodium channels, shifting the voltage still further. The inflow of sodium ions accelerates until the inner surface of the membrane is locally positive. The voltage reversal closes the sodium channel and opens the potassium channel. The outflow of potassium ions quickly restores the negative potential. The voltage reversal, known as the action potential, propagates itself down the axon (1, 2). After a brief refractory period a second impulse can follow (3). The impulse-propagation speed is that measured in the giant axon of the squid.

nylate cyclase. Adenylate cyclase and related membrane enzymes are known to serve a number of regulatory functions in neurons, and the precise mechanisms of these actions are now under active investigation.

In the course of the embryonic development of the nervous system a cell must be able to recognize other cells so that the growth of each cell will proceed in the right direction and give rise to the right connections. The process of cell-cell recognition and the maintenance of the structure arrived at by such recognition depend on special classes of membrane proteins that are associated with unusual carbohydrates. The study of the protein-carbohydrate complexes associated with cell recognition is still at an early stage.

The intrinsic membrane proteins I have been describing are neither distributed uniformly over the cell surface nor all present in equal amounts in each neuron. The density and the type of protein are governed by the needs of the cell and differ among types of neuron and from one region of a neuron to another. Thus the density of channels of a particular type ranges from zero up to about 10,000 per square micrometer. Axons generally have no chemically gated channels, whereas in postsynaptic membranes the density of such channels is limited only by the packing of the channel molecules. Similarly, dendritic membranes typically have few voltage-gated channels, whereas in axon membranes the density can reach 1,000 channels per square micrometer in certain locations.

The intrinsic membrane proteins are synthesized primarily in the body of the neuron and are stored in the membrane in small vesicles. Neurons have a special transport system for moving such vesicles from their site of synthesis to their site of function. The transport system seems to move the vesicles along in small jumps with the aid of contractile proteins. On reaching their destination the proteins are inserted into the surface membrane, where they function until they are removed and degraded within the cell. Precisely how the cell decides where to put which membrane protein is not known. Equally unknown is the mechanism that regulates the synthesis, insertion and destruction of the membrane proteins. The metabolism of membrane proteins constitutes one of cell biology's central problems.

How do the properties of the various membrane proteins I have been discussing relate to neuron function? To approach this question let us now return to the nerve impulse and examine more closely the molecular properties that underlie its triggering and propagation. As we have seen, the interior of the neuron is about 70 millivolts negative with respect to the exterior. This "resting po-

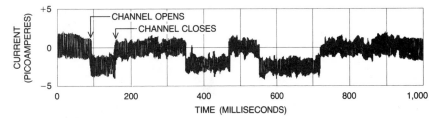

RESPONSE OF A SINGLE MEMBRANE CHANNEL to the transmitter compound acetylcholine is revealed by a recently developed technique that has been applied by Erwin Neher and Joseph H. Steinbach of the Yale University School of Medicine. Acetylcholine-activated channels, which are present in postsynaptic membranes, allow the passage of roughly equal numbers of sodium and potassium ions. The record shows the flow of current through a single channel in the postsynaptic membrane of a frog muscle activated by the compound suberyldicholine, which mimics action of acetylcholine but keeps channels open longer. Experiment shows that channels open on an all-or-none basis and stay open for random lengths of time.

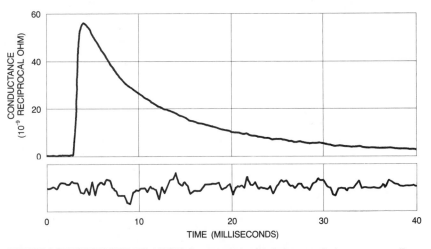

SODIUM CHANNELS IN AN AXON also operate in simple open-or-shut manner as well as independently of one another, according to investigations conducted by Frederick J. Sigworth of the Yale University School of Medicine. During the propagation of a nerve impulse about 10,000 channels normally open in a myelin-free region of the axon membrane, namely a node of Ranvier. The upper trace depicts the sodium permeability at such a node as a function of time. The lower trace, recorded at a 12-fold amplification of the upper one, shows fluctuations in permeability around the average due to the random opening and closing of channels.

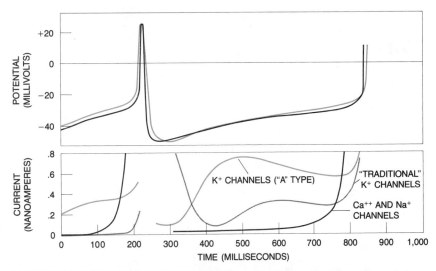

NERVE IMPULSES IN BODIES OF NEURONS require the coordinated opening and closing of five types of channel permeable to various kinds of ion (sodium, potassium or calcium). The contribution of the different channels to the nerve impulse can be represented by simultaneous nonlinear differential equations. The upper pair of curves represent an actual recording of voltage changes as a function of time in the body of a neuron (black) and changes computed from equations (color). The lower curves depict the current carried by the principal types of channel as a function of time. A complicated interaction of channel types is required to achieve a train of nerve impulses. The study on which curves are based was carried out by John A. Connor at the University of Illinois and by the author at the Yale University School of Medicine.

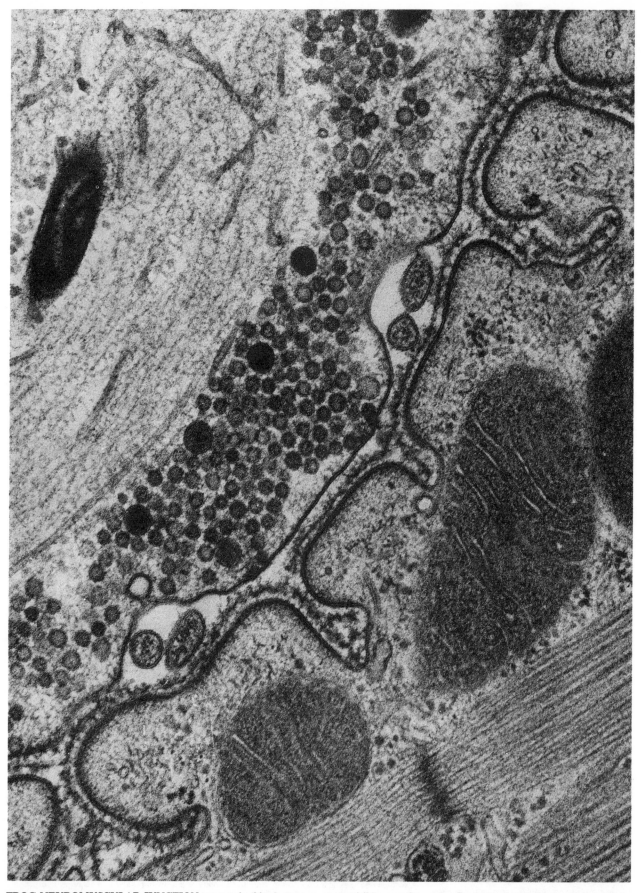

FROG NEUROMUSCULAR JUNCTION appears in this electron micrograph made by Heuser. The synaptic cleft separates the axon at the upper left from the muscle cell at the lower right. Synaptic vesicles cluster along the presynaptic membrane, with two synaptic con- **tacts visible near the center. Postsynaptic membrane of the muscle cell exhibits a feature that is not seen at other synapses: the membrane forms postjunctional folds opposite each contact. Freeze-fracture replicas of presynaptic membrane are shown on opposite page.**

tential" is a consequence of the ionic disequilibrium brought about by the sodium pump and by the presence in the cell membrane of a class of permanently open channels selectively permeable to potassium ions. The pump ejects sodium ions in exchange for potassium ions, making the inside of the cell about 10 times richer in potassium ions than the outside. The potassium channels in the membrane allow the potassium ions immediately adjacent to the membrane to flow outward quite freely. The permeability of the membrane to sodium ions is low in the resting condition, so that there is almost no counterflow of sodium ions from the exterior to the interior even though the external medium is tenfold richer in sodium ions than the internal medium. The potassium flow therefore gives rise to a net deficit of positive charges on the inner surface of the cell membrane and an excess of positive charges on the outer surface. The result is the voltage difference of 70 millivolts, with the interior being negative.

The propagation of the nerve impulse depends on the presence in the neuron membrane of voltage-gated sodium channels whose opening and closing is responsible for the action potential. What are the characteristics of these important channel molecules? Although the sodium channel has not yet been well characterized chemically, it is a protein with a molecular weight probably in the range of 250,000 to 300,000 daltons. The pore of the channel measures about .4 by .6 nanometer, a space through which sodium ions can pass in association with a water molecule. The channel has many charged groups critically placed on its surface. These charges give the channel a large electric dipole moment that varies in direction and magnitude when the molecular conformation of the channel changes as the channel goes from a closed state to an open one.

Because the surface membrane of the cell is so thin the difference of 70 millivolts across the resting membrane gives rise to a large electric field, on the order of 100 kilovolts per centimeter. In the same way that magnetic dipoles tend to align themselves with the lines of force in a magnetic field, the electric dipoles in the sodium-channel protein tend to align themselves with the membrane electric field. Changes in the strength of the membrane field can therefore drive the channel from the closed conformation to the open one. As the inner surface of the membrane is made more positive by the entering vanguard of sodium ions the sodium channels tend to spend an increasing fraction of their time in the open conformation. The process in which the channels are opened by a change in the membrane voltage is known as sodium-channel activation.

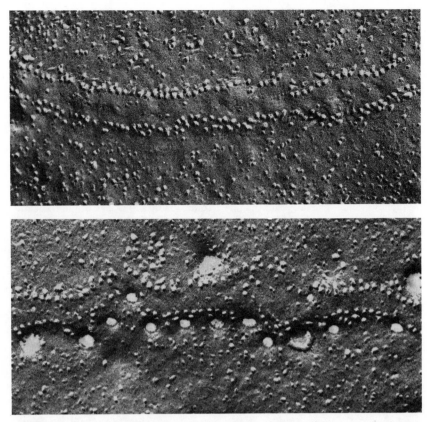

FREEZE-FRACTURE REPLICAS of the presynaptic membrane of the frog neuromuscular junction were made by Heuser. The upper micrograph shows the membrane three milliseconds after the muscle had been stimulated. Running across the axon membrane is a double row of particles: membrane proteins that may be calcium channels or structural proteins to which vesicles attach. The lower micrograph shows the membrane five milliseconds after stimulation. The stimulation has caused synaptic vesicles to fuse with presynaptic membrane and form pits.

The process is terminated by a phenomenon called sodium inactivation. Voltage differences across the membrane that cause sodium channels to open also drive them into a special closed conformation different from the conformation characteristic of the channel's resting state. The second closed conformation, called the inactivated state, develops more slowly than the activation process, so that channels remain open briefly before they are closed by inactivation. The channels remain in the inactivated state for some milliseconds and then return to the normal resting state.

The complete cycle of activation and inactivation normally involves the opening and closing of thousands of sodium channels. How can one tell whether the increase in overall membrane permeability reflects the opening and closing of a number of channels in an all-or-none manner or whether it reflects the operation of channels that have individually graded permeabilities? The question has been partly answered by a new technique that relates fluctuations in membrane permeability to the inherently probabilistic nature of conformational changes in the channel proteins. One can trigger repeated episodes of channel opening and calculate the average permeability at a particular time and also the exact permeability on a given trial. The exact permeability fluctuates 10 percent or so around a mean value. Analysis of the fluctuations shows that the sodium channels open in an all-or-none manner and that each channel opening increases the conductance of the membrane by 8×10^{-12} reciprocal ohms. One of the principal challenges in understanding the neuron is the development of a complete theory that will describe the behavior of the sodium channels and relate it to the molecular structure of the channel protein.

As I noted briefly above, axons also have voltage-gated potassium channels that help to terminate the nerve impulse by letting potassium ions flow out of the axon, thereby counteracting the inward flow of sodium ions. In the cell body of the neuron the situation is still more complex, because there the membrane is traversed by five types of channel. The different channels open at different rates, stay open for various intervals and are preferentially permeable to different species of ions (sodium, potassium or calcium).

The presence of the five types of channel in the cell body of the neuron, compared with only two in the axon, gives rise to a more complex mode of nerve-impulse generation. If an axon is presented with a maintained stimulus, it generates only a single impulse at the onset of the stimulus. Cell bodies, however, generate a train of impulses with a frequency that reflects the intensity of the stimulus.

Neurons are able to generate nerve impulses over a wide range of frequencies, from one or fewer per second to several hundred per second. All nerve impulses have the same amplitude, so that the information they carry is represented by the number of impulses generated per unit of time, a system known as frequency coding. The larger the magnitude of the stimulus to be conveyed, the faster the rate of firing.

When a nerve impulse has traveled the length of the axon and has arrived at a terminal button, one of a variety of transmitters is released from the presynaptic membrane. The transmitter diffuses to the postsynaptic membrane, where it induces the opening of chemically gated channels. Ions flowing through the open channels bring about the voltage changes known as postsynaptic potentials.

Most of what is known about synaptic mechanisms comes from experiments on a particular synapse: the neuromuscular junction that controls the contraction of muscles in the frog. The axon of the frog neuron runs for several hundred micrometers along the surface of the muscle cell, making several hundred synaptic contacts spaced about a micrometer apart. At each presynaptic region the characteristic synaptic vesicles can be recognized readily.

Each of the synaptic vesicles contains some 10,000 molecules of the transmitter acetylcholine. When a nerve impulse reaches the synapse, a train of events is set in motion that culminates in the fusion of a vesicle with the presynaptic membrane and the resulting release of acetylcholine into the cleft between the presynaptic and the postsynaptic membranes, a process termed exocytosis. The fused vesicle is subsequently reclaimed from the presynaptic membrane and is quickly refilled with acetylcholine for future release.

Many details of the events leading to exocytosis have recently been elucidated. The fusion of vesicles to the presynaptic membrane is evidently triggered by a rapid but transient increase in the concentration of calcium in the terminal button of the axon. The arrival of a nerve impulse at the terminal opens calcium channels that are voltage-gated and allows calcium to flow into the terminal. The subsequent rise in calcium concentration is brief, however, because the terminal contains a special apparatus that rapidly sequesters free calcium and returns its concentration to the normal very low level. The brief spike in the free-calcium level leads to the fusion of transmitter-filled vesicles with the presynaptic membrane, but the precise mechanism of this important process is not yet known.

Interesting details of the structure of the terminal membrane have been revealed by the freeze-fracture technique, a method that splits the layers of the bilayer membrane and exposes the intrinsic membrane proteins for examination by electron microscopy. In the frog neuromuscular junction a double row of large membrane proteins runs the width of each synapse. Synaptic vesicles become attached on or near the proteins. Only these vesicles then fuse to the membrane and release their transmitter; other vesicles seem to be held in reserve some distance away. The fusion of vesicles is a random process and occurs independently for each vesicle.

In less than 100 microseconds acetylcholine released from fused vesicles diffuses across the synaptic cleft and binds to the acetylcholine receptor: an intrinsic membrane protein embedded in the postsynaptic membrane. The receptor is also a channel protein that is chemically gated by the presence of acetylcholine. When two acetylcholine molecules attach themselves to the channel, they lower the energy state of the open conformation of the protein and thereby increase the probability that the channel will open. The open state of the channel is a random event with an average lifetime of about a millisecond. Each packet of 10,000 acetylcholine molecules effects the opening of some 2,000 channels.

During the brief period that a channel is open about 20,000 sodium ions and a roughly equal number of potassium ions pass through it. As a result of this ionic flow the voltage difference between the two sides of the membrane tends to approach zero. How close it approaches to zero depends on how many channels open and how long they stay open. The acetylcholine released by a typical nerve impulse produces a postsynaptic potential, or voltage change, that lasts for only about five milliseconds. Because postsynaptic potentials are produced by chemically gated chan-

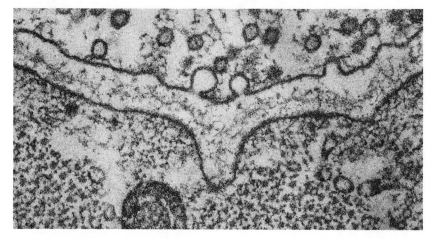

TRANSMITTER IS DISCHARGED into the synaptic cleft at the synaptic junctions between neurons by vesicles that open up after they fuse with the axon's presynaptic membrane, a process called exocytosis. This electron micrograph made by Heuser has caught the vesicles in the terminal of an axon in the act of discharging acetylcholine into the neuromuscular junction of a frog. The structures that appear in the micrograph are enlarged some 115,000 diameters.

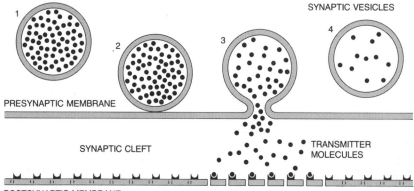

SYNAPTIC VESICLES are clustered near the presynaptic membrane. The diagram shows the probable steps in exocytosis. Filled vesicles move up to synaptic cleft, fuse with the membrane, discharge their contents and are reclaimed, re-formed and refilled with transmitter.

nels rather than by voltage-gated ones they have properties quite different from those of the nerve impulse. They are usually smaller in amplitude, longer in duration and graded in size depending on the quantity of transmitter released and hence on the number of channels that open.

Different types of chemically gated channels exhibit different selectivities. Some resemble the acetylcholine channel, which passes sodium and potassium ions with little selectivity. Others are highly selective. The voltage change that results at a particular synapse depends on the selectivity of the channels that are opened. If positive ions move into the cell, the voltage change is in the positive direction. Such positive-going voltage channels tend to open voltage-gated channels and to generate nerve impulses, and so they are known as excitatory postsynaptic potentials. If positive ions (usually potassium) move out of the cell, the voltage change is in the negative direction, which tends to close voltage-gated channels. Such postsynaptic potentials oppose the production of nerve impulses, and so they are termed inhibitory. Excitatory and inhibitory postsynaptic potentials are both common in the brain.

Brain synapses differ from neuromuscular-junction synapses in several ways. Whereas at the neuromuscular junction the action of acetylcholine is always excitatory, in the brain the action of the same substance is excitatory at some synapses and inhibitory at others. And whereas acetylcholine is the usual transmitter at neuromuscular junctions, the brain synapses have channels gated by a large variety of transmitters. A particular synaptic ending, however, releases only one type of transmitter, and channels gated by that transmitter are present in the corresponding postsynaptic membrane. In contrast with neuromuscular channels activated by acetylcholine, which stay open for about a millisecond, some types of brain synapses have channels that stay open for less than a millisecond and others have channels that remain open for hundreds of milliseconds. A final major difference is that whereas the axon makes hundreds of synaptic contacts with the muscle cell at the frog's neuromuscular junction, axons in the brain usually make only one or two synaptic contacts on a given neuron. As might be expected, such different functional properties are correlated with significant differences in structure.

As we have seen, the intensity of a stimulus is coded in the frequency of nerve impulses. Decoding at the synapse is accomplished by two processes: temporal summation and spatial summation. In temporal summation each postsynaptic potential adds to the cumulative total of its predecessors to yield a voltage change whose average amplitude reflects the frequency of incoming nerve impulses. In other words, a neuron that is firing rapidly releases more transmitter molecules at its terminal junctions than a neuron that is firing less rapidly. The more transmitter molecules that are released in a given time, the more channels that are opened in the postsynaptic membrane and therefore the larger the postsynaptic potential is. Spatial summation is an equivalent process except that it reflects the integration of nerve impulses arriving from all the neurons that may be in synaptic contact with a given neuron. The grand voltage change derived by temporal and spatial summation is encoded as nerve-impulse frequency for transmission to other cells "downstream" in the nerve network.

I have described what is usually regarded as the normal flow of information in neural circuits, in which postsynaptic voltage changes are encoded as nerve-impulse frequency and transmitted over the axon to other neurons. In recent years, however, a number of instances have been discovered where a postsynaptic potential is not converted into a nerve impulse. For example, the voltage change due to a postsynaptic potential can directly cause the release of transmitter from a neighboring site that lacks a nerve impulse. Such direct influences are thought to come into play in synapses between dendrites and also in certain reciprocal circuits where one dendrite makes a synaptic contact on a second dendrite, which in turn makes a synaptic contact back on the first dendrite. Such direct feedback seems to be quite common in the brain, but its implications for information processing remain to be worked out.

Much current investigation of the neuron focuses on the membrane proteins that endow the cell's bilayer membrane, which is otherwise featureless, with the special properties brain function depends on. With regard to channel proteins there are many unanswered questions about the mechanisms of gating, selectivity and regulation. Within the next five or 10 years it should be possible to relate the physical processes of gating and selectivity to the molecular structure of the channels. The basis of channel regulation is less well understood but is now coming under intensive investigation. It seems that hormones and other substances play a role in channel regulation that is now becoming appreciated. The central problems at synaptic junctions involve exocytosis and other activities related to the metabolism and release of transmitters. One can expect increasing attention to be focused on the role of the surface membrane in the growth and development of neurons and their synaptic connections, the remarkable process that establishes the integration of the nervous system.

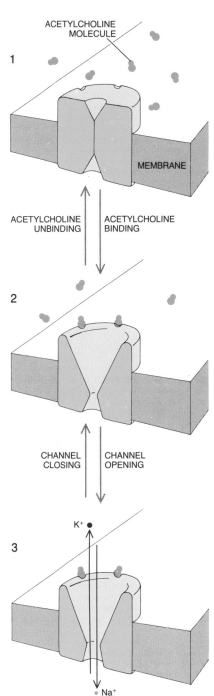

ACETYLCHOLINE CHANNEL in a postsynaptic membrane is opened by acetylcholine molecules' discharging into the synaptic cleft. The drawing shows the acetylcholine receptor at the frog neuromuscular junction. Two acetylcholine molecules bind rapidly to the resting closed channel to form a receptor-acetylcholine complex (1, 2). The complex undergoes a change in its conformation that opens the channel to the passage of sodium and potassium ions (3). The time required for conformational change in the complex limits the speed of the reaction. The channel remains open for about a millisecond on the average and then reverts to the receptor-acetylcholine complex. While it is open the channel passes about 20,000 sodium ions and an equal number of potassium ions. The acetylcholine rapidly dissociates and is destroyed by the enzyme acetylcholine esterase. Acetylcholine receptor appears in mircograph on page 19.

2

Calcium in Synaptic Transmission

by Rodolfo R. Llinás
October 1982

A current of calcium ions triggers the passage of signals from one nerve cell to another. The process is studied in a synapse (a neuronal junction) hundreds of times a synapse's usual size

Synapses are the sites at which neurons, or nerve cells, communicate with one another. At an earlier stage in the life of an organism they are important in determining how the nervous system develops. (Synapses form at the tips of the fibers that sprout from the body of a neuron.) It seems certain that much of the brain's ability to regenerate after injury and much of an organism's ability to learn will ultimately be explained in terms of the function of synapses. Furthermore, it is becoming clear that most of the diseases of the brain and many psychiatric disorders result from a disruption of synaptic communication or are associated with such a disruption. The synapse is the weakest link in brain activity: it is the first part of a neuronal chain to be fatigued by a high level of message transmission, and it is the site of action for most of the drugs that affect the brain, including addictive substances as well as therapeutic ones from aspirin to barbiturates. For all these reasons a detailed understanding of every aspect of synaptic transmission is essential if we are to understand how the brain works and how the brain can malfunction.

Here I shall be concerned in particular with the aspect of transmission called depolarization-release coupling. In synaptic transmission one neuron (the presynaptic cell) releases a biologically active substance (a neurotransmitter), which evokes a response, either excitatory or inhibitory, in a second neuron (the postsynaptic cell). The release of the transmitter is in essence a process of secretion, a process shared by cells throughout the evolutionary sequence. In virtually every known instance secretion is accomplished by exocytosis, a mechanism in which vesicles, or membranous sacs, inside the cell fuse with the membrane that surrounds the cell. The fusion everts the contents of the vesicles into the extracellular environment.

A neuron evidently releases neurotransmitter in the same way. The release is known to be stimulated when the membrane of the presynaptic neuron at the synapse loses its electrical polarization. The step in synaptic transmission that concerns me here can be expressed, then, by a question: How does the depolarization of the membrane lead to the release of neurotransmitter, so that the neuron can act on the next cell in the neuronal chain? It turns out that the connection between the electrical activity of the cell and the release of neurotransmitter is not direct; an essential intermediary is the calcium ion.

When a neuron is at rest, there is an electric potential, or voltage difference, of some 70 millivolts between the inside of the cell and the outside. The voltage is negative inside, and so the cell membrane is said to have a polarization of −70 millivolts. When the membrane is depolarized, the voltage difference diminishes. Indeed, in the type of depolar-

ization called the action potential a reverse potential develops with a value of from +10 to +30 millivolts. The reversal persists for only about a millisecond.

The precise mechanism by which depolarization arises is inconsequential as far as the subsequent release of neurotransmitter is concerned. In neurons such as sensory receptors that respond to stimuli (a burst of sound, a touch on the skin) impinging directly on the surface of the cell, the depolarization is caused by the energy of the stimulus itself. In such a neuron the depolarization can be a subtle change in potential that lasts for many seconds. In neurons in the brain the depolarization is generally caused by the signals each cell receives at the synapses it makes with other neurons. Here the depolarization is typically a rapid modulation in voltage that lasts for only a few milliseconds. If it is sufficiently great, it can induce the cell to generate its own action potential. In ev-

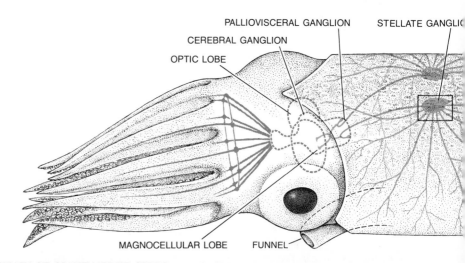

CHAIN OF GIANT NERVE CELLS on each side of the midline in the squid *Loligo pealii* includes a giant synapse: a site of signaling from one nerve cell to another that is .7 millimeter long, or several hundred times the size of a typical synapse. It is so large that it can be impaled by several microelectrodes and still transmit signals while its potential (that is, its voltage) is recorded or even varied experimentally. Here the chain of nerve cells that includes the synapse is shown in color. The first cell, the first-order giant neuron (*green*), is in the magnocellular lobe of what amounts to the brain of the animal. The neuron gathers sensory data from organs such

ery case, however, the effect of a wave of depolarization arriving at the membrane of the presynaptic terminal of a neuron is the same: it causes transmitter to be released.

Depolarization is nonetheless not sufficient in itself to cause the release of transmitter. In addition a supply of calcium ions must be present in the extracellular environment. Depolarization seems to be the means by which an inward current of calcium ions is induced to flow through the membrane of the presynaptic terminal. In this respect too synaptic transmission resembles other known secretory processes. In every secretory process in which the test has been made secretion is triggered by an increase in the concentration of calcium inside the secretory cell. The increase is the result of the entry of calcium from the external environment or of the release of calcium from internal stores. It seems likely that in neurons the entering calcium ions promote the fusion of special intracellular vesicles (synaptic vesicles) into the presynaptic membrane. The vesicles are stationed near the inner surface of the membrane at the site of transmitter release and are filled with the transmitter substance.

Among the difficulties that arise in attempts to advance our understanding beyond these basic points, one problem is implacable: synapses are small. So far the only measuring technique fast enough and precise enough to serve in the study of events in a functioning synapse is the recording of its electrical activity. Yet the diameter of a presynaptic terminal in most vertebrate and invertebrate species is typically from .1 micrometer to five micrometers. The micro-electrode with which one can probe the synapse's electrical properties has a diameter of about .5 micrometer. Piercing the presynaptic terminal without damaging the synapse is quite difficult, and so is keeping the electrode in place as the experiment proceeds.

Fortunately for the experimenter some synapses are considerably larger than the average. They include certain synapses found in ganglia (clusters of neurons) in mollusks and crustaceans and the synapse between the neurons called Mauthner cells in the brain stem of fishes and some amphibians. The Mauthner cells govern the movement of the tail. Among large synapses, however, the most notable example is a giant synapse of the squid. It is some 700 micrometers long. Because of its size and accessibility it has yielded most of what is now known directly about the relation of depolarization to the release of neurotransmitter.

The presence of a giant synapse in the squid was first reported in 1935 by J. Z. Young of the Marine Biological Association of the United Kingdom at Plymouth. The synapse was included in his description of a chain of giant nerve cells in the squid; the chain consists of three axons, or nerve fibers, and the neuronal cell bodies from which they arise. The axons' diameters are so great that some of them had been mistaken for blood vessels; Young identified them as nervous tissue on the basis of their ability to conduct an action potential. The chains on each side of the squid's body are now known to govern the animal's flight response and capture of prey.

The first nerve cell in the chain is called the first-order giant neuron. It is in the magnocellular lobe, an assemblage of neurons in what amounts to the animal's brain. Its cell body is some 150 micrometers in diameter, which makes it many times larger than the largest nerve cells in the human brain. With its extensive set of dendrites (tubular extensions of the cell body) it is 800 micrometers long. By means of these dendrites it gathers signals that originate in such places as the eyes, the vestibular organs and the tentacles. It is in effect a single-cell computer that assesses danger to the animal. The results of its calculations are transmitted along the membrane of its axon. As if to ensure the synchrony of such signals, the axons of the first-order giant neurons on each side of the body fuse for a short distance at the midline.

The second cell in the chain, the second-order giant neuron, is in the palliovisceral ganglion behind the magnocellular lobe. The second-order cell is 100 micrometers in diameter. Its axon curves through the ganglion on a trajectory that directs it toward the squid's mantle. Along the curve it comes in contact with the axon of the first-order giant cell, which establishes a synapse with it. Then the second-order axon continues on to the center of the stellate ganglion, on the inner surface of the mantle. There it broadens and divides into a set of from eight to 10 terminal branches that look like fingers radiating from the palm of a hand. The branches vary in size: the thinnest are 25 micrometers in diameter and the thickest are twice as big.

The third cell in the chain is actually a set of hundreds of cells, each of which is about 50 micrometers in diameter. The

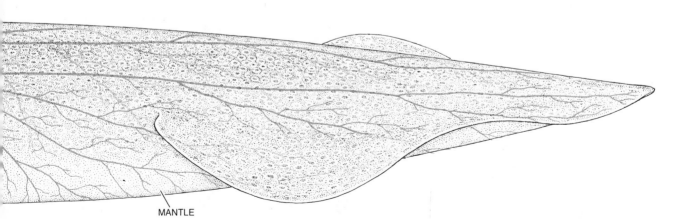

MANTLE

as the eyes and the tentacles, and it dispatches neural signals along its axon (nerve fiber) to the palliovisceral ganglion. There the axon synaptically contacts the axon emitted by the second cell in the chain, the second-order giant neuron (*red*). In the stellate ganglion at the flank of the animal's mantle the second-order axon synaptically contacts a set of third-order axons (*blue*) that trigger the mantle's contraction. The contraction sends water jetting out of the animal's fun-nel, propelling the animal from danger. The synapse between the second-order axon and the largest third-order axon, the one passing almost rearward through the muscle tissue of the mantle near the midline, is the giant synapse studied by the author and his colleagues. The squid is shown about 1.5 times life size; the stellate ganglion on the left side of the body (the ganglion enclosed in this drawing by a rectangle) is further enlarged in the top illustration on the next page.

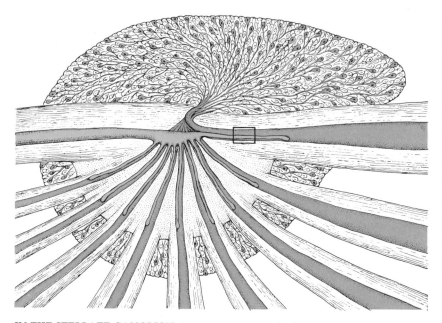

IN THE STELLATE GANGLION the second-order axon (*red*) branches into a series of fin-gerlike extensions. Each one is a presynaptic terminal: it makes synaptic contact with a third-order axon (*blue*), which arises from the fusion of the axons emitted by several hundred neu-rons in the ganglion. The largest third-order axon, shown leaving the ganglion toward the right at the center of a bundle of smaller axons, forms the postsynaptic part of the giant synapse. Be-yond the synapse the axon widens to .5 millimeter, or a hundred times the diameter of a typical axon. The part of the giant synapse enclosed by the rectangle is enlarged in the illustration below.

duces a synchronous excitation in all the third-order axons. The speed with which a wave of depolarization propa-gates along a nerve fiber depends on the diameter of the fiber; since the longest of the third-order axons are also the thickest, the synchronization of the sig-nals is preserved and all the muscles of the mantle contract simultaneously. The contraction forces the water in the mantle out through the funnel near the head of the animal. Thus the squid can escape from danger by jet propulsion.

The giant synapse we have studied is the one between the second-order axon and the largest third-order axon. When the synapse is examined in detail, it is found that the membrane of the presyn-aptic terminal (the second-order axon) is smooth. In contrast, the membrane of the postsynaptic terminal (the third-order axon) has multiple branchings that divide repeatedly and end as a net of thorn-shaped extensions. In the giant synapse as many as 5,000 such exten-sions face the presynaptic membrane. This structure is unusual. In a typi-cal synapse the postsynaptic terminal is smooth, whereas the presynaptic termi-nal consists of protuberances at the end of an axon that are called synaptic bou-tons (from the French word for button).

Nevertheless, a microscopic examina-tion of the places in the giant synapse where the presynaptic membrane abuts a postsynaptic thorn shows morphology quite typical of a synapse. Inside the pre-synaptic terminal at such places there are synaptic vesicles and two other in-tracellular structures that are notably abundant at synapses: mitochondria, the organelles that make energy available to the cell, and subcysternal systems, which consist of folds of intracellular membrane. (The function of the latter is not known.) In the same areas the post-synaptic thorn has a thickened mem-brane, which is recognized in the typical synapse to be the site where receptors are found. The receptors are molecules embedded in the postsynaptic mem-brane that react with arriving molecules of neurotransmitter.

cells are organized into from eight to 10 sets, and in each set all the axons fuse to form a single giant axon. Hence the stel-late ganglion gives rise to from eight to 10 giant axons. The thinnest of them is about 50 micrometers in diameter; it re-ceives signals through a synapse with the thinnest of the terminal branches of the second-order axon, then it widens somewhat and proceeds away from the stellate ganglion to innervate muscle tissue in the mantle near the ganglion. The thickest of the third-order axons is about 200 micrometers in diameter; it

receives signals through a synapse with the thickest of the terminal branches of the second-order axon, then it widens to a diameter of about 500 micrometers and extends to innervate muscle tissue in the most distant part of the mantle. The other third-order axons have inter-mediate destinations.

The rudiments of how this network functions can be inferred from its struc-ture. When stimuli reaching the first-order neuron exceed some threshold that signifies alarm, the cell activates the second-order axon, which in turn pro-

It was in the giant synapse of the squid that it was first demonstrated directly that the transmission of a neural signal across the synapse requires the release of a transmitter substance. This inter-mediate chemical step (first proposed in the course of research on the union between nerve and muscle) was estab-lished in the 1950's and the early 1960's through a number of studies by Theo-dore H. Bullock, Susumu Hagiwara, Ichiji Tasaki, Noriko and Akira Take-uchi, Ricardo Miledi and C. R. Slater. The studies capitalized on the size of the giant synapse: the workers were able to position the tip of a microelectrode in the presynaptic terminal and the tip of a second microelectrode in the postsynap-

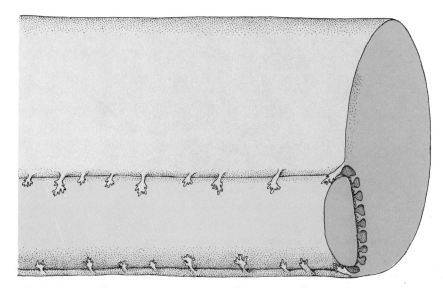

GIANT SYNAPSE is made up of two apposed membranes. The postsynaptic membrane of the third-order axon has some 5,000 spiny extensions. They face the presynaptic membrane.

tic axon so that the voltage across the membrane of each terminal could be recorded simultaneously. In addition the workers could vary the concentration of ions or drugs in the solution in which they had isolated the synapse (along with a few centimeters of the second-order and third-order axons). Thus they could study the role of the ions and the action of the drugs.

It emerged from the various studies that when an action potential propagating along the second-order axon reaches the presynaptic terminal, it causes a depolarization of the postsynaptic axon after a delay of about a millisecond. On the other hand, an action potential at the postsynaptic terminal (induced by electrical stimulation of the third-order axon) has no effect on the presynaptic terminal. It further emerged that the degree of postsynaptic depolarization increases with the amplitude and the duration of the presynaptic action potential. Yet the presynaptic action potential becomes incapable of causing postsynaptic depolarization if all the calcium is removed from the solution in which the synapse is bathed. The combination of all these findings (showing, among other things, synaptic delay and unidirectionality) was strong evidence that signals are not transmitted electrically across the giant synapse. They are carried by a chemical messenger.

In 1966 Bernard Katz and Miledi, working at the Zoological Station in Naples, and two groups working at the Marine Biological Laboratory in Woods Hole, Mass., one group consisting of Kiyoshi Kusano, D. R. Livengood and Robert Werman and the other of James R. Bloedel, Peter W. Gage, David M. J. Quastel and me, made a further discovery. Signal transmission across the giant synapse does not require the arrival of a presynaptic action potential. It requires only that the presynaptic terminal be depolarized, which can be accomplished not only by an action potential but also by passing an electric current into the terminal through a microelectrode that impales it. The discovery was significant because it separated the mechanism of electrical excitability responsible for the action potential from the mechanism responsible for the release of transmitter from the presynaptic terminal.

The amount of transmitter released by the presynaptic terminal under artificial stimulation turned out to depend in a curious way on the degree of presynaptic depolarization. A slight depolarization of the presynaptic membrane causes a small postsynaptic depolarization; hence one infers that a relatively small amount of transmitter is released from the presynaptic terminal. A larger presynaptic depolarization leads to a larger response, but only up to a point: the response decreases if the presynaptic depolarization is greater than 60 milli-

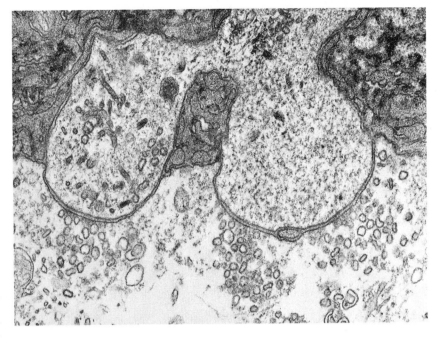

ELECTRON MICROGRAPH of the giant synapse suggests how it transmits signals. The presynaptic terminal of the synapse is at the bottom; it is abutted by two bulbous protrusions arising from a single spiny extension of the postsynaptic axon. Inside the presynaptic terminal at the places of abutment are accumulations of saclike organelles. They are synaptic vesicles, which are thought to contain a neurotransmitter: a substance released by the presynaptic terminal that acts on the membrane of the postsynaptic terminal. The signaling is thus a secretory process. The electron micrograph was made at an enlargement of 74,000 diameters by David W. Pumplin of the University of Maryland at Baltimore School of Medicine and Thomas S. Reese of the National Institute of Neurological and Communicative Disorders and Stroke.

volts with respect to its resting value of −70 millivolts, or in other words if the presynaptic potential attains a value more positive than −10 millivolts. If the presynaptic depolarization is great enough, so that the presynaptic potential becomes greater than +100 millivolts, the postsynaptic response can actually disappear. Even then, however, brief synaptic transmission is detected at the end of the experimental pulse as the presynaptic potential returns to its resting value.

These observations were interpreted in terms of the effect of depolarization on the flow of calcium ions. When the neuron is in its resting state, the membrane is impermeable to calcium ions, which have a much higher concentration outside the cell than inside. Depolarization evidently opens channels in the membrane that allow calcium ions to pass; as a result the ions flow inward, propelled by the electromotive force arising from what remains of the electric potential across the membrane as well as the lesser concentration of calcium inside the terminal. If the depolarization is sufficiently strong, however, the influx of ions is opposed by electrical forces. The calcium ion carries two positive electric charges, and when the interior of the cell becomes electrically positive, the ions are repelled. Thus even though the membrane channels are

opened by the depolarization, the calcium ions cannot enter the cell and neurotransmitter is not released. The degree of presynaptic depolarization at which synaptic transmission is abolished was named the suppression potential, and the postsynaptic response at the end of a pulse was named the off postsynaptic potential.

The early results of the study of the giant synapse lent support, then, to the hypothesis, first proposed by Katz, that calcium triggers the release of neurotransmitter. Still, the calcium hypothesis came of age only when the results of three further lines of investigation became known in the late 1960's and early 1970's.

First, Katz and Miledi demonstrated that calcium ions are capable of generating action potentials, but only at the presynaptic terminal, and that these action potentials are accompanied by the release of transmitter. In particular they demonstrated that if the ability of the presynaptic membrane to pass sodium and potassium ions is blocked by drugs, action potentials can nonetheless be detected along with a subsequent postsynaptic depolarization. What modulates the voltage across the membrane of a neuron as an action potential propagates is the redistribution of ions and hence of electric charge between the in-

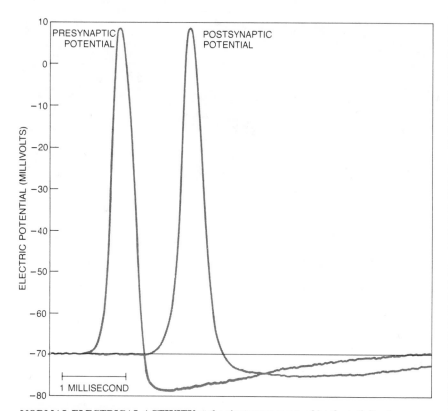

NORMAL ELECTRICAL ACTIVITY at the giant synapse resembles the activity at synapses in all vertebrate and invertebrate animals. When the electric potential across the membrane of the presynaptic terminal is altered from its resting value of −70 millivolts by a spike of voltage called an action potential, which is transmitted along the length of the second-order axon, the presynaptic terminal releases neurotransmitter. The arrival of the transmitter at the postsynaptic terminal alters the permeability of the postsynaptic membrane to ions. The resulting ionic current through the membrane stimulates the postsynaptic terminal to develop a voltage spike.

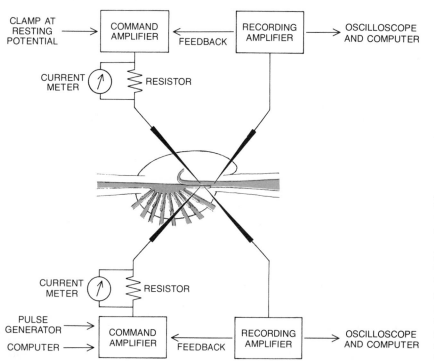

VOLTAGE-CLAMP CIRCUITRY imposes a predetermined pattern of voltage on the presynaptic terminal or the postsynaptic terminal or both of them. One microelectrode monitors the voltage across the membrane of the presynaptic terminal. Its measurements serve as feedback for a second microelectrode, which injects current into the terminal. Instruments monitor the amount of current injected to give the terminal a "clamped" voltage, set by a pulse generator or a computer. A similar method clamps the postsynaptic terminal at its resting potential.

terior of the cell and the extracellular environment. Sodium ions flow in; then potassium ions flow out. The action potentials Katz and Miledi detected must have depended entirely on the passage of calcium ions. From their results they inferred that the presynaptic membrane has calcium channels. The action potentials were abolished when calcium was removed from the bathing solution.

Second, Miledi did a much simpler experiment. He showed that in the absence of extracellular calcium the mere injection of calcium ions into the presynaptic terminal causes a postsynaptic depolarization. Therefore the injected calcium must have caused transmitter to be released.

The third line of work was done by John R. Blinks, Charles Nicholson and me. We showed that if the presynaptic terminal is filled with aequorin, a protein that emits light when it is exposed to calcium, light can be detected in the presynaptic terminal during normal synaptic transmission. It follows that the concentration of calcium in the terminal increases at such times. A further experiment by Nicholson and me showed that the light signals synaptic events quite faithfully. For example, when we applied a suppression potential to the presynaptic terminal, we failed to detect light during the depolarizing pulse. We did detect light when we stopped the pulse and the postsynaptic terminal responded with an off potential. This demonstrated that the suppression of transmission is indeed due specifically to a suppression of the entry of calcium ions.

Beyond all doubt, then, calcium is the agent that triggers synaptic transmission. Two points remained to be clarified: the relation between depolarization and the entry of calcium and the subsequent relation between the entry of calcium and the release of transmitter. With this in mind my colleagues and I decided to make a change in our experimental technique. In essence we had been using a current clamp: we had injected into the presynaptic terminal a constant electric current and measured the resulting presynaptic depolarization (the changing voltage across the membrane of the terminal). The membrane has the electrical properties of resistance and capacitance, that is, it resists the flow of electric current and also stores electric charge, causing a delay in its response to a current pulse. In general, therefore, the procedure of current clamping resembles the charging of a capacitor in parallel with a resistor and the measurement of the voltage across the pair.

The difficulty with the current-clamping method is that the electrical properties of the membrane of a neuron are more complex than those of a capacitor in parallel with a resistor. Specifically,

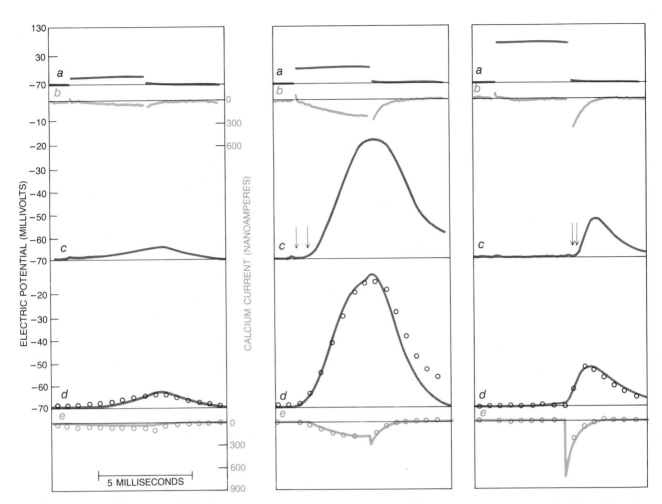

VOLTAGE-CLAMP EXPERIMENTS done by the author and his colleagues subjected the presynaptic terminal of the giant synapse to an artificial "step" in voltage (*a*). By means of drugs the terminal's membrane was made impermeable to all ions except calcium; thus the current across the membrane (*b*) represented the flow of calcium ions only. The current was measured by monitoring the current the clamp circuitry injected to counteract the change in voltage that the calcium current causes. In the graph at the left the voltage step is 30 millivolts. A small calcium current flows slowly into the presynaptic terminal. The calcium influx causes neurotransmitter to be released, a fact that can be inferred from the change in the potential of the postsynaptic terminal (*c*). In the middle graph the voltage step is 52 milli-volts. The calcium current increases faster and builds to a higher level; the postsynaptic response is also more pronounced, although it does not begin until nearly a millisecond (*arrows*) after the onset of the voltage step. In the graph at the right the voltage step is 130 millivolts. The calcium current is suppressed and so is the postsynaptic response. At the end of the voltage step, however, a "tail current" of calcium flows. The postsynaptic terminal responds in only .2 millisecond (*arrows*). The data collected in voltage-clamp experiments yield a model of synaptic transmission: a mathematical description of how the synapse responds to voltage. At the bottom of each panel the postsynaptic potential (*d*) and the presynaptic calcium current (*e*) resulting from the model are compared with the experimental data (*dots*).

the changing voltage across the membrane can open channels that allow various kinds of ions (including calcium ions) to enter and leave the neuron. In short, the membrane's conductivity to ions is voltage-dependent. The movements of ions (which constitute an electric current) cause further changes in the voltage, which cause still further changes in the membrane, and so on. The most extreme result of this membrane self-modulation is an action potential. Learning the properties of the membrane when it is actively changing its voltage and no longer responding passively to experimental manipulation is problematic. Yet the active properties of the membrane are central to a nerve cell's functioning.

We determined, therefore, to end our work with current-clamping and take the opposite tack: we would employ the voltage-clamp technique that Kenneth S. Cole had devised at the Marine Biological Laboratory. A. L. Hodgkin and A. F. Huxley of the University of Cambridge had employed the technique to elucidate (in the third-order giant axon of the squid) the electrical events underlying the action potential. In the voltage-clamp technique a burst of current drives the membrane of a neuron very rapidly to a given level of depolarization. The level is then kept constant ("clamped") in spite of the movement of ions through channels that have opened in the membrane. The clamping is accomplished by an electronic feedback circuit that varies the continuing injection of current.

The voltage-clamp technique has two advantages. First, the amount of current injected when the clamp is in effect must exactly balance the current of ions; hence the metering of the applied current constitutes a measurement of the ionic current. It is a measurement that cannot otherwise be made, since the microelectrodes record only voltages. Second, the step in voltage that marks the onset of the clamping is much faster than the changes it induces in the structure of the membrane. This means the clamp circuitry charges the membrane's capacitance well before ionic currents have started to flow through membrane channels. It follows that the ionic currents can be distinguished from the injected current that merely gets stored in the terminal; therefore we could hope

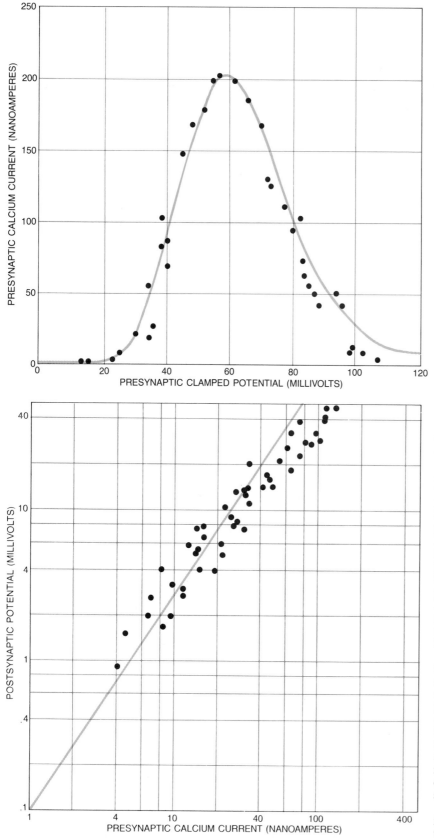

in our further work with the giant synapse to study the full time course of changes in the presynaptic membrane's conductivity (or equivalently the opening and closing of channels and the passage of ions through them) for a given steady level of depolarization. Indeed, we could hope to derive from measurements of the current of calcium ions at various levels of clamped depolarization a mathematical model of the presynaptic calcium conductance, that is, a mathematical expression that would suggest how the membrane's calcium channels change in response to a changing pattern of voltage.

During the summers of 1975 through 1978 Kerry Walton and I did a series of voltage-clamp experiments at the Marine Biological Laboratory. We blocked the conductance of the presynaptic membrane to sodium and potassium, so that our measurements would represent only the flow of calcium ions. We found to our delight that the calcium current was readily measured. It turns out that the membrane's conductivity to calcium depends in a complex way on the level of clamped depolarization.

Consider the time course of the current. In response to a small step of depolarizing voltage the calcium current increases quite slowly: it reaches a plateau only after several milliseconds. In response to a larger voltage step it attains its plateau value in a fraction of a millisecond. Consider also the amplitude of the plateau. Almost no calcium current is detected unless the voltage step is at least 15 millivolts. For a larger step the plateau amplitude is markedly greater. For a step of 60 millivolts (which gives the terminal membrane a clamped potential of −10 millivolts) the amplitude is maximal. For still larger steps the amplitude decreases, a result that Katz and Miledi's discovery of the suppression potential would lead one to expect. For steps greater than 140 millivolts the current is again quite small.

At the end of an episode of clamping the presynaptic potential falls from its clamped value back to the resting value of −70 millivolts. The return to resting potential (which can be accomplished quickly by the voltage-clamp circuitry) is accompanied by a second and distinctive flow of calcium ions into the presynaptic terminal. It is called the tail current, and it has been observed for ions other than calcium. It has no delay: it starts as soon as the voltage step ends.

The tail current can be explained as follows. Throughout the voltage step the calcium channels in the presynaptic membrane are open, and at the end of the step they close. The closing of the channels, however, is slower than the return of the potential to its resting value. The channels therefore remain open briefly after the potential has decayed,

RESULTS OF VOLTAGE-CLAMP EXPERIMENTS are summarized in these graphs. The relation between the clamped value of the presynaptic voltage and the plateau level of the calcium current that flows into the presynaptic terminal (*top graph*) is nonlinear: the current has its maximum for a clamped voltage of about 60 millivolts, and it is negligible for clamping above about 110 millivolts. On the other hand, the relation between the calcium current and the postsynaptic response (*bottom graph*) is quite linear. Both axes in the bottom graph are logarithmic.

at a time when the electromotive force again is favorable for the entry of calcium ions.

Doubtless the tail current is responsible for the release of transmitter that causes the off postsynaptic response. It is remarkable, then, that the off response begins a mere .2 millisecond after the tail current starts to flow. It is all the more remarkable because the normal postsynaptic response to the arrival of an action potential at the presynaptic terminal is delayed by as much as a millisecond. One must conclude that a large part of the synaptic delay is due to the time required for the calcium channels to open when the presynaptic membrane is first depolarized. The subsequent events in synaptic transmission must be fast indeed. In particular, once calcium ions have entered the presynaptic terminal they must act quite rapidly to get transmitter released.

The next goal of our work was to devise from our data a mathematical description relating the calcium current to any given pattern of voltage. To this end we began a collaboration with Izchak Z. Steinberg of the Weizmann Institute of Science in Israel. The heart of such a mathematical model is a description of how the calcium channels in the membrane respond to depolarization. The hypotheses with which Steinberg, Walton and I began were similar to those proposed by Hodgkin and Huxley for sodium and potassium channels. We assumed that a driving force arises from the difference between the concentration of calcium ions inside the presynaptic terminal and the concentration outside it. We further assumed that an increase in the presynaptic membrane's conductivity to calcium ions can be triggered by voltage-dependent changes in the structure of calcium channels that cause them to open. One of the simplest possibilities is that each channel consists of a certain number of subunits and that each subunit has two possible states, s and s'. Probably each state corresponds to a different shape. If all the subunits in a channel are in the state s' (the activated state), the channel is open; otherwise it is closed. The probability that a subunit will enter state s' is determined by the voltage across the membrane; the greater the extent of depolarization is, the more likely the subunit is to be activated. One further assumption is that the subunits do not interact with one another but that each subunit changes its shape independently.

From these assumptions (together with the laws of thermodynamics) it is possible to construct a family of equations expressing the time course of the calcium current in response to a clamped presynaptic voltage. The current is proportional to the rate at which

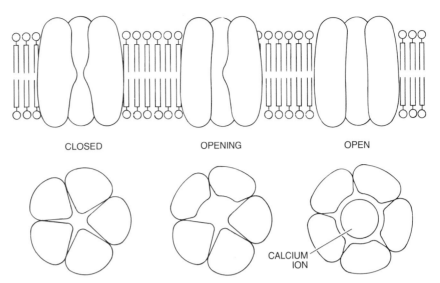

CALCIUM CHANNELS in the membrane of the presynaptic terminal must open in response to a change in voltage in order for the calcium current to flow. The model of synaptic transmission devised by the author and his colleagues suggests that five independent changes in the structure of each channel are needed to open it. Hence each channel may hypothetically be represented by a rosette consisting of five proteins each of which extends through the membrane. The voltage must change the shape of all five proteins before calcium ions can enter. The membrane of a neuron and its terminals consists itself of two layers of lipids (fatty molecules).

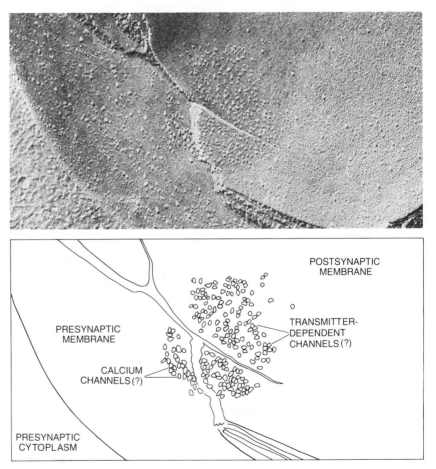

SMALL PARTICLES in the membrane of the presynaptic terminal may be the calcium channels. Particles in the postsynaptic membrane may be channels that open in response to the arrival of transmitter molecules. The tissue shown in this scanning electron micrograph was frozen, then cracked. The cracking characteristically splits a neuronal membrane down the middle (that is, between its lipid layers). Each particle is thus embedded in the membrane. The electron micrograph was made by Pumplin and Reese at an enlargement of 105,000 diameters.

ions flow through an open gate multiplied by the fraction of the gates that are open. Each equation incorporates a different assumption about the number of subunits per gate. In the equation that best matches the time course and voltage-dependence of the calcium current we had measured in our voltage-clamp experiments the number of subunits is five.

One hypothetical configuration for the channel (a configuration offered in the absence of any direct evidence) is a set of five identical proteins, each one spanning the thickness of the presynaptic membrane; the five protein molecules might be arranged in the form of a rosette. When all the proteins have the shape s', they circumscribe a channel that allows the entry of calcium ions. Recently, however, the study of individual calcium channels by several investigators has suggested an alternative configuration in which each channel is circumscribed by a helix of proteins.

The voltage-clamp experiments show

that the calcium current saturates, or attains a maximum value. Hence each channel must be able to pass only a certain number of calcium ions per unit of time regardless of the driving force. To explain the saturation we assume that each channel has an energy barrier. Perhaps a part of each protein molecule juts slightly into the channel and has a positive electric charge. Because the charge would tend to repel an entering calcium ion the driving force would have to "push" ions through the channel and their rate of flow would have an upper limit. The voltage-clamp experiments also show that the falloff in calcium current at the end of a voltage step is simply an exponential decrease and not the more gradual falloff that would indicate a complex sequence of events. This finding fits the hypothesis that a change in the shape of only one subunit in an open channel suffices to close it.

With a mathematical model for the response of the calcium channels to a given clamped level of voltage we could

calculate the response of the channels to an arriving action potential. We approximated the action potential by a series of incremental changes in voltage and then we employed the model to predict the calcium current that would result from the succession of increments. We do not know the number of calcium channels in the presynaptic membrane, nor do we know the conductivity of an individual channel, and so the model's results could suggest not the actual magnitude of the calcium current but only its time course and its relative magnitude. Nevertheless, it appeared from the results that the calcium current starts to flow at the end of the action potential as the presynaptic voltage returns to its resting level.

Our third goal was to relate the presynaptic inflow of calcium to the release of neurotransmitter. Since the delay between the two events is only some .2 millisecond, we hypothesized that the calcium enters the presynaptic terminal quite near the site where transmitter is released. We further supposed the latter site would be marked in electron micrographs by the presence of synaptic vesicles.

Electron microscopy of the giant synapse does indeed show "active zones." They are characterized by a gathering of vesicles in the presynaptic terminal and a thickening in the membrane of the postsynaptic terminal. Scanning electron micrographs made by the freeze-fracture method reveal additional structure. In the freeze-fracture technique a section of tissue including the giant synapse is frozen and then cracked. The tissue tends to cleave down the middle of either the presynaptic membrane or the postsynaptic one. Under the electron microscope the freeze-fracture specimens show that the presynaptic membrane at an active zone has embedded in it as many as 1,500 small particles. We hypothesize (with other investigators) that they are the sites of calcium entry.

In order to incorporate into our model the relation of calcium entry to transmitter release we made some further assumptions. They are again the simplest possibilities. In the presynaptic terminal we assume that calcium binds to a molecule we call the fusion-promoting factor. In response to the binding a certain part of the factor molecule enters an active state by way of a first-order kinetic reaction (a reaction whose overall rate depends only on the concentration of a single reactant, in this case the fusion-promoting factor itself).

The activated fusion-promoting factor causes synaptic vesicles to fuse with the presynaptic membrane so that they release their content of neurotransmitter. The rate of the reaction depends only on the concentration of the activated factor. Then the activated factor re-

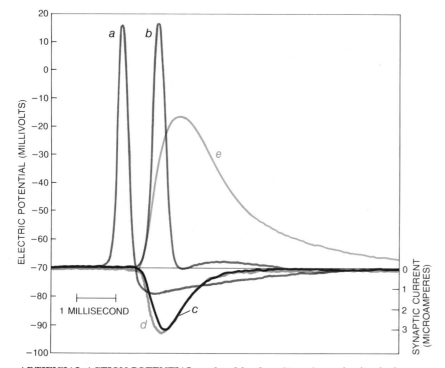

ARTIFICIAL ACTION POTENTIAL produced by the voltage-clamp circuitry is shown to have effects on the giant synapse identical with those of a natural action potential. First a natural action potential (*a*) is recorded by a microelectrode positioned in the presynaptic terminal. Normally it produces a postsynaptic action potential (*b*). Now, however, the postsynaptic terminal is voltage-clamped at its resting potential. Under this circumstance the transmitter molecules arriving at the postsynaptic terminal open channels in the membrane and ions flow through them, but the clamp circuitry injects current in precisely the right amount to keep the net current zero and the voltage unaltered. The amount of current the clamp injects (*c*) is therefore equal in time course and magnitude to the "synaptic current" through the postsynaptic membrane. Next the membrane channels on both sides of the synapse that open in response to a change in voltage (as opposed to the arrival of transmitter) are incapacitated by drugs, so that they no longer pass sodium and potassium ions, the basis of a natural action potential. If the presynaptic voltage-clamp circuitry delivers the voltage pattern of the action potential recorded earlier, the resulting synaptic current (*d*) has the same latency, amplitude and time course as the one caused by natural events. If the postsynaptic terminal is then unclamped but the drugs remain, the postsynaptic terminal responds to the artificial action potential by showing a prolonged change in voltage (*e*). The change represents the passage of ions through transmitter-dependent channels. Ordinarily the postsynaptic change in voltage serves to open the voltage-dependent channels in the membrane, so that a postsynaptic action potential is generated.

turns to an inactive state. Meanwhile the neurotransmitter molecules are opening channels in the postsynaptic membrane so that ionic currents flow and the membrane becomes depolarized. The model resulting from these assumptions proved capable of reproducing the postsynaptic responses we had measured in the voltage-clamp experiments, including the postsynaptic off potential at the end of a presynaptic depolarization exceeding the suppression potential.

Thus far our voltage-clamp studies had yielded a set of experimental results amenable to a mathematical description of the presynaptic calcium current and its relation to the amplitude of the postsynaptic potential. Ultimately, however, the model must be tested against actual measurements of the calcium current during the course of an action potential. In 1979 and 1980 the test was attempted. In a series of experiments done by Mutsuyuki Sugimori, Sanford M. Simon and me at the Marine Biological Laboratory a presynaptic action potential and the resulting postsynaptic action potential were recorded simultaneously in the giant synapse. The recorded changes in membrane potential were stored in the memory of a digital computer. We then employed the recorded signals in experiments of a new kind.

As in the earlier voltage-clamp experiments the voltage-dependent conductance of the membranes for sodium and potassium was blocked chemically, leaving the presynaptic membrane with only its voltage-dependent conductance for calcium and leaving the postsynaptic membrane with only the ionic channels that are opened by the arrival of transmitter molecules. The new experiments differed in that we no longer clamped the presynaptic terminal at a constant level of depolarization. Instead the amplifier and the microelectrode that inject current into the presynaptic terminal were driven by the presynaptic voltage spike we had recorded earlier. In this way we artificially imposed on the terminal the voltage pattern of an action potential even though the normal basis of the action potential (the inward flow of sodium ions and the outward flow of potassium ions) was absent. The postsynaptic response to the artificial action potential was virtually identical with the response to the natural voltage spike; hence the artificial potential caused the release of transmitter from the presynaptic terminal with an identical amplitude and delay.

We felt justified, therefore, in thinking that further measurements made when ionic conduction was suppressed and the presynaptic terminal was stimulated by an artificial action potential would be a valid representation of natural events. Accordingly, we monitored the amount of current the voltage-clamp circuitry injected into the presynaptic terminal

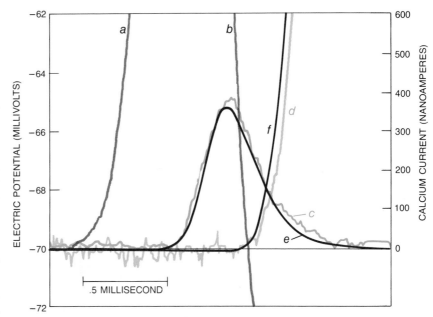

ROLE OF CALCIUM in synaptic transmission can be determined by means of the artificial action potential because the artificial voltage spike can be applied while the flow of various ions across the presynaptic membrane is being blocked pharmacologically. First the artificial spike is applied while sodium and potassium are being blocked. On the scale of the graph only the upshoot (a) and fall (b) of the spike are visible. The amount of current the clamp circuit must supply to produce the spike is monitored; it represents the sum of the amount that changes the voltage across the membrane in the absence of flows of sodium and potassium and the amount that precisely offsets the unblocked flow of calcium. The experiment is repeated with calcium also blocked. The difference between the two results corresponds to the calcium current alone (c). One concludes that an action potential causes the release of neurotransmitter from the presynaptic terminal by causing a current of calcium ions to enter the terminal during the falling phase of the voltage spike. The postsynaptic response to the transmitter (d) begins soon after. The graph also shows the calcium current (e) and the postsynaptic response (f) predicted by the model of synaptic transmission devised by the author and his colleagues.

to make the artificial spike. Since sodium and potassium currents were being blocked, the injected current had two components. Part of it depolarized the membrane and thereby made up for the absence of the sodium and potassium currents; the rest exactly counterbalanced the calcium current that was flowing unblocked through the membrane. Next we chemically blocked the calcium current and repeated the experiment. The difference between the two results is the calcium current alone. It compares closely with the calcium current predicted by our model.

Quite recently our results have been confirmed by the work of Steven J. Smith, Milton P. Charlton and Robert S. Zucker at the Marine Biological Laboratory and that of Miledi and Parker in Naples. The work employed a dye that changes color in the presence of calcium. The dye is injected into the presynaptic terminal of the giant synapse. Its change in color there suggests a time course for the calcium current much like the one we measured by voltage-clamping.

Many prospects are now before us. For some years I have been tantalized by the possibility that synaptic transmission is a modified form of neuronal

growth. For one thing, the concentration of calcium in developing neural tissue seems to control the rate at which a growth cone, which is the growing tip of an axon, adds new cell membrane to its surface. Then too the membrane of a growth cone has a voltage-dependent conductance for calcium ions.

Perhaps the synaptic terminal can be regarded as a modified growth cone in which growth has been subdued, so that there is no longer any permanent increase in the area of the membrane. In an embryonic neuron vesicles in a growth cone might fuse with the growth-cone membrane to increase the extent of the membrane and give it newly minted proteins. Later in the life of the organism the vesicles in the presynaptic terminal that arises from the growth cone would fuse with the membrane of the terminal, evert their content of neurotransmitter and get taken back into the terminal. (Such recycling was proposed some years ago by John E. Heuser and Thomas S. Reese.) The mechanism that serves growth and plasticity in the developing nervous system would then come to serve neuronal signaling. The understanding of the molecular processes underlying this sequence of events holds the highest promise in the search for the ways of the brain.

II

CHEMISTRY OF
THE BRAIN

CHEMISTRY OF THE BRAIN

II

INTRODUCTION

Investigations of the chemistry of the brain, particularly of synaptic transmission, may answer such fundamental questions about the brain and behavior as how it stores memories, why sex is such a powerful motivation, and what the biologic basis of mental illness is. Identifying chemical circuits in the brain is a difficult task, but here too progress has been made, as in the detection of circuits that may be at fault in Alzheimer's and Parkinson's disease. The study of chemical transmitter substances and circuits in the brain, and the subsequent drug actions on these circuits and neurons, began only a few years ago, but has greatly expanded to become the largest field of neuroscience.

The first chemicals to be identified as synaptic transmitter agents, such as acetylcholine, were relatively small, simple molecules. In the past few years, however, a number of peptide substances (and receptors) have been discovered—many are hormones produced either by the pituitary gland or body tissues and organs, but some are present in nerve cells and are thought to function as transmitter chemicals. A peptide, incidentally, is simply a relatively short chain of amino acids. Proteins are much larger and more complex combinations of amino acids. The body manufactures some amino acids, but there are about 9—the essential amino acids—that must be obtained from protein foods. Perhaps the most interesting of the brain's peptides are the recently discovered brain opioids, substances present in the pituitary gland and brain that act like opium. Progress has been rapid in this area due to the extent of our knowledge of the pharmacology of morphine and related opiates.

Because synaptic transmission is fundamentally chemical, the brain must be seen broadly as a complex chemical system requiring the addition of many chemical substances to function properly; hence the critical importance of nutrition. This can be illustrated by the critically important transmitter substance, acetylcholine. How much acetylcholine is manufactured by nerve cells is determined by the amount of choline we consume in food.

The first article in this section is a broad treatment of the chemistry of the brain and synaptic transmitters by Leslie Iverson of Maudsley Hospital in London. The article by Floyd Bloom of the Salk Institute that follows focusses on the neuropeptides. These peptides are thought to function as transmitter chemicals and hormones in the nervous system. In the third article, Richard Wurtman of the Massachusetts Institute of Technology investigates the key topic of nutrition and the brain.

3

The Chemistry of the Brain

by Leslie L. Iversen
September 1979

Signals are sent from one neuron to another by diverse chemical transmitters. These chemical systems, overlaid on the neuronal circuits of the brain, add another dimension to brain function

Neurons share the biochemical machinery of all other living cells, including the ability to generate chemical energy from the oxidation of foodstuff and to repair and maintain themselves. Among the specialized features they possess and other cells do not are those that have to do with the special function of neurons as transmitters of nerve impulses, such as their need to maintain ionic gradients, involving a high rate of energy consumption, and those associated with the ability of neurons to manufacture and release a special array of chemical messengers known as neurotransmitters. At synapses, the microscopic regions of close proximity between the terminal of one neuron and the receiving surface of another, the arrival of an impulse causes a sudden release of molecules of transmitter from the terminal. The transmitter molecules then diffuse across the fluid-filled gap between the two cells and act on specific receptor sites in the postsynaptic membrane, thereby altering the electrical activity of the receiving neuron.

Some 30 different substances are known or suspected to be transmitters in the brain, and each has a characteristic excitatory or inhibitory effect on neurons. The transmitters are not randomly distributed throughout the brain but are localized in specific clusters of neurons whose axons project to other highly specific brain regions. The superimposition of these diverse chemically coded systems on the neuronal circuitry endows the brain with an extra dimension of modulation and specificity.

Considerable progress has been made in recent years in characterizing the various transmitter substances (although many more undoubtedly remain to be discovered), in mapping their distribution in the brain and in elucidating the molecular events of synaptic transmission. Such research has revealed that the behavioral effects of many drugs and neurotoxins arise from their ability to disrupt or modify chemical transmission between neurons. It has also hinted that the causes of mental illness may ultimately be traced to defects in the functioning of specific transmitter systems in the brain.

In terms of general energy metabolism the brain is the most active energy consumer of all body organs, a fact reflected in its large blood supply and oxygen uptake. Although the human brain represents only 2 percent of the total body weight, its rate of oxygen utilization (50 milliliters per minute) accounts for 20 percent of the total resting utilization of oxygen. This enormous expenditure of energy is thought to be due to the need to maintain the ionic gradients across the neuronal membrane on which the conduction of impulses in the billions of brain neurons depends. Moreover, there is no respite from this energy demand: the rate of brain metabolism is relatively constant day and night and may even increase somewhat during the dreaming phases of sleep. To put matters in perspective, however, the total energy equivalent of brain metabolism is only some 20 watts.

An important recent advance in studies of energy metabolism in the brain is the development by Louis Sokoloff and his colleagues at the National Institute of Mental Health of a method that makes it possible to visualize the rate of energy metabolism in brain cells. Neurons adjust the rate at which they take up glucose to fulfill their metabolic needs at the time. Hence active neurons take up glucose more rapidly than quiescent ones. The glucose taken up is normally metabolized rapidly, but a chemical analogue of glucose, 2-deoxyglucose, is taken up into the cells by the same uptake mechanism but is not metabolized there. If radioactively labeled deoxyglucose is injected into the bloodstream, it will accumulate in neurons, and the rate at which it accumulates can serve as an indicator of the cells' metabolic activity. The accumulation of radioactive deoxyglucose can be seen and measured by placing thin sections of frozen brain on radiation-sensitive film. Areas that are rich in labeled material show up when the film is developed. The technique has opened an entirely new realm in brain research, since it makes it possible to detect what cells in the brain were active during a given experimental procedure. For example, the precise areas of the brain receiving visual inputs from the left or right eye can be visualized by flashing a light into one eye or the other.

Whereas the other body organs are able to utilize a variety of alternative fuels (such as sugars, fats and amino acids), neurons can utilize only blood glucose. Moreover, whereas tissues such as muscle are able to function for short periods in the absence of oxygen, the brain is entirely dependent on oxidative metabolism. If the supply of oxygenated blood to the brain is interrupted, consciousness is lost within 10 seconds and permanent damage to the brain ensues. Similar effects result from any condition that lowers blood glucose, such as when a diabetic injects himself with an overdose of insulin. Although elaborate control mechanisms ensure that the blood pressure will remain stable and that

NEURONS CONTAINING NOREPINEPHRINE, a chemical transmitter in the brain, glow brilliantly in this section of rat brain viewed in the fluorescence microscope. The norepinephrine-containing cells, situated in a region of the brain stem called the locus coeruleus, were made visible by reacting them with glyoxylic acid, which converts norepinephrine into a fluorescent chemical derivative. Thousands of other neurons are also present in the field, but because they contain other transmitters they are not visible. The norepinephrine neurons in the locus coeruleus project their axons to many parts of the brain, including the cerebellum and the forebrain. They are thought to be involved in sleep, mood and brain reward. Micrograph was made by Floyd E. Bloom, Gary S. Jones and Jacqueline F. McGinty of the Salk Institute.

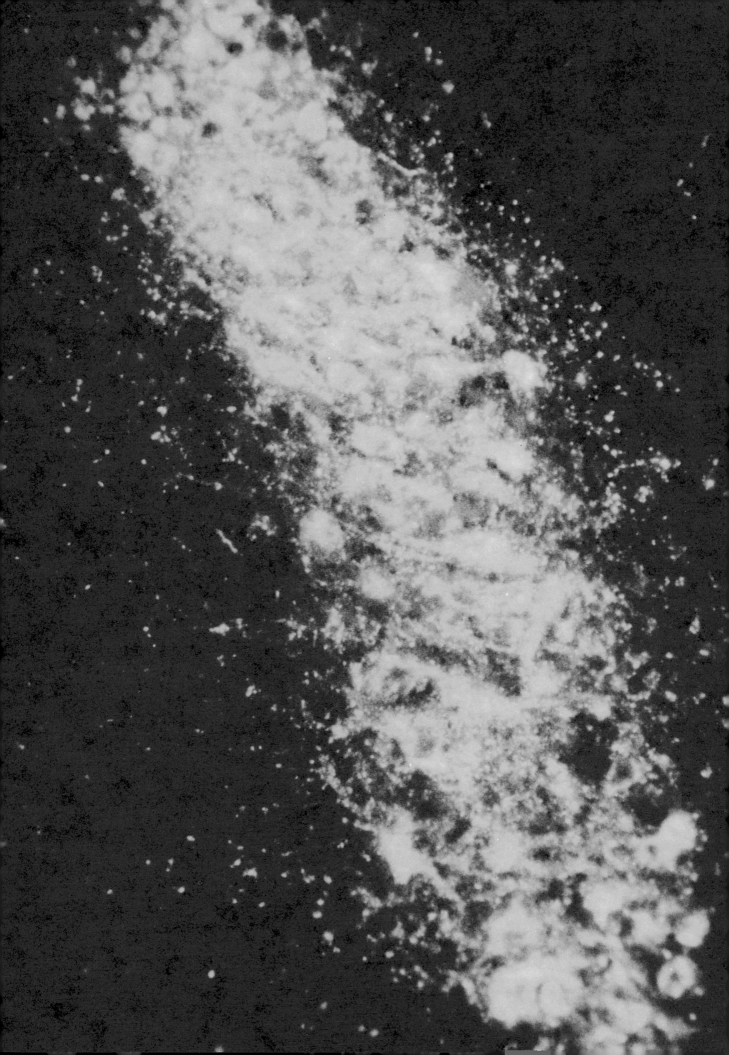

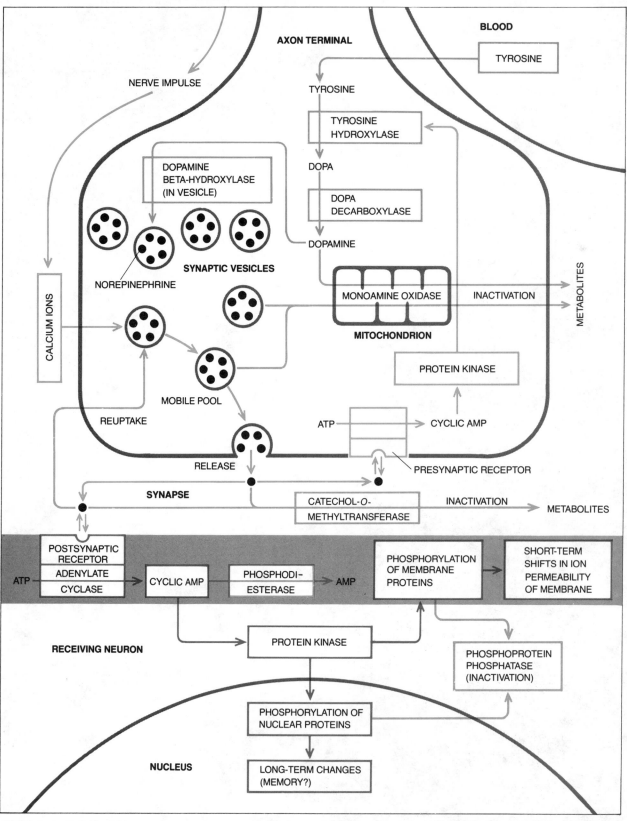

CHEMICAL TRANSMISSION across the synapse, the narrow gap between two neurons in the brain, involves an elaborate sequence of molecular events. Here the process of transmission at a norepinephrine synapse is diagrammed. First norepinephrine is manufactured from the amino acid tyrosine in three steps, each of which is catalyzed by an enzyme. The transmitter is then stored within membrane-bound vesicles in association with storage proteins (*indicated by green pathways*). The arrival in the axon terminal of a nerve impulse triggers an influx of calcium ions, which induces the release of norepinephrine from the vesicles into the synaptic space (*red pathways*).

The liberated transmitter molecules bind to specific receptor proteins embedded in the postsynaptic membrane, triggering a series of reactions that culminate in short-term (electrical) and long-term effects on the receiving neuron (*purple pathways*). The action of norepinephrine is then terminated by a variety of means, including rapid reuptake of the transmitter into the axon terminal and degradation by enzymes (*blue pathways*). The release of some norepinephrine into the synaptic space activates presynaptic receptors on the axon terminal, initiating production of cyclic AMP, which activates protein kinase, thus stimulating more norepinephrine production (*orange pathway*).

there will be constant levels of oxygen and glucose in the blood, it seems clear that the enormous behavioral flexibility made possible by the expanded size and capacity of the mammalian brain has been acquired during evolution at a high metabolic cost.

As cells go, neurons are exceedingly sensitive: their function can be disrupted by toxic substances that find their way into the bloodstream and also by small molecules that are normally present in the blood, such as amino acids. This sensitivity may explain why the brain is isolated from the general circulation by the selective filtration system known as the blood-brain barrier. The effectiveness of the barrier is due to the relative impermeability of the blood vessels in the brain and to the presence of tight sheaths of glial cells (the supporting cells of the brain) around the blood vessels. Although small molecules such as those of oxygen can pass readily through the barrier, most of the larger molecules required by brain cells, such as those of glucose, must be actively taken up by special transport mechanisms. The blood-brain barrier has important consequences for the design of drugs that act directly on the brain: if such substances are to cross the barrier, their molecules must be either very small or readily soluble in the fatty membranes of the glial cells. A few select regions of the brain are not shielded by the blood-brain barrier; they include structures that are specifically responsive to blood-borne hormones or whose job it is to monitor the chemical composition of the blood.

Within individual neurons other transport problems are presented by the fact that part of the cell is represented by extended thin fibers. The axon that carries the nerve impulse away from the cell body of a neuron can be millimeters or centimeters long. The neurons of the adult brain cannot be replaced and must last a lifetime, so that there must be mechanisms to renew all their components. This requirement calls for the synthesis by the cell of enzymes and other complex molecules, and such synthesis can proceed only in the region of the cell nucleus, that is, in the cell body of the neuron. Therefore replacing the components of the axon requires a means of transporting components substantial distances within the cell. Indeed, there is a constant movement of proteins and other components from the cell body down the entire length of the axon.

This phenomenon of axonal transport was discovered more than 30 years ago by Paul A. Weiss and his colleagues at the University of Chicago. Until that time it was generally assumed that the axoplasm—the jellylike fluid inside the axon—was merely an inert mechanical support for the excitable membrane that propagated the nerve impulse. When

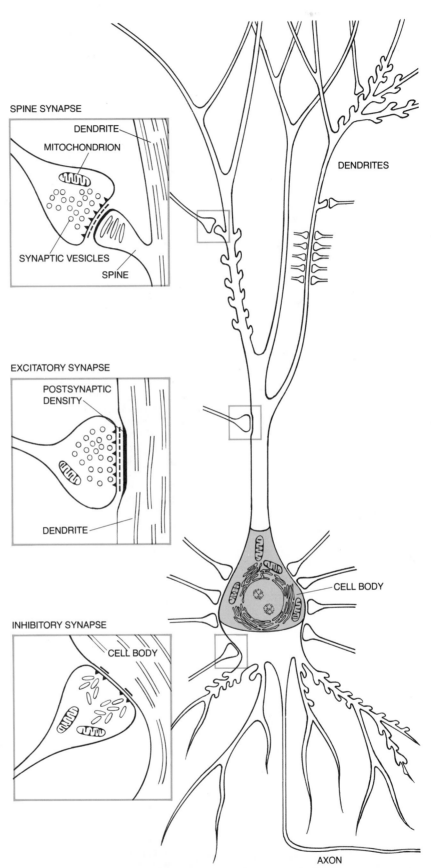

SYNAPSES impinging on a typical neuron in the brain are either excitatory or inhibitory, depending on the type of transmitter that is released. Such synapses can be distinguished morphologically in the electron microscope: excitatory synapses tend to have round vesicles and a continuous dense thickening of the postsynaptic membrane, and inhibitory synapses tend to have flattened vesicles and a discontinuous postsynaptic density. Synapses can also be classified according to their position on the surface of the receiving neuron. They may be on the cell body, on the trunk of the dendrites, on "spines" projecting from the dendrites or on the axon.

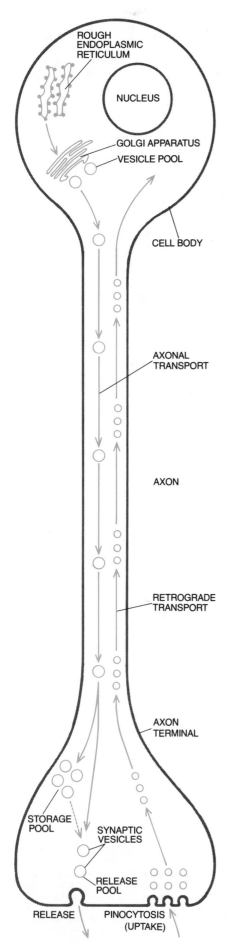

ROUGH
ENDOPLASMIC
RETICULUM

NUCLEUS

GOLGI APPARATUS
VESICLE POOL

CELL BODY

AXONAL
TRANSPORT

AXON

RETROGRADE
TRANSPORT

AXON
TERMINAL

STORAGE
POOL

SYNAPTIC
VESICLES

RELEASE
POOL

RELEASE   PINOCYTOSIS
(UPTAKE)

Weiss constricted the axon at a given point, however, he found that after several days the fiber bulged out on the side of the constriction toward the cell body and narrowed on the side away from the cell body. When he released the constriction, the dammed-up axoplasm resumed its flow.

It is now known that the axoplasm is an artery for a busy molecular traffic moving in both directions between the cell body of the neuron and its axon terminals. There are several different systems involved, including a slow-transport system in which material flows away from the cell body at a speed of about a millimeter a day, and a faster-transport system in which material flows in both directions at speeds of between 10 and 20 centimeters a day. The slow-transport system represents a bulk flow of axoplasm carrying components important for the growth and regeneration of the axon; the faster-transport system represents the flow of more specialized cellular components, including some of the enzymes involved in the manufacture of transmitters.

It is not yet understood how such differential rates of transport are achieved, but both the slow and the fast mechanisms seem to involve the numerous fibrous proteins in the axon that are visible in electron micrographs. By following the transport of radioactively labeled proteins outward along axons it has been possible to trace the precise anatomical connections between neurons in the brain. The connections between the terminals of a neuron in one region of the brain and the cell body in another region can also be mapped by means of the enzyme horseradish peroxidase, which has the special ability to travel rapidly up the axon in a retrograde fashion.

The functional chemistry of the brain is exceedingly difficult to study. Not only are the transmitter substances present in vanishingly small quantities but also the brain tissue is structurally and chemically so complex that it is not easy to isolate a given transmitter system for examination. New techniques have been developed, however, to overcome these formidable obstacles. One major advance came in the early 1960's with the

AXONAL TRANSPORT is responsible for moving cellular components such as vesicles and enzymes from their site of manufacture in the neuronal cell body to the axon terminals, which may be millimeters or centimeters away. Transport in the retrograde direction from the axon terminals to the cell body carries factors essential to health of the neuron. Axonal transport can be exploited to trace pathways by observing the movement along axons of radioactively labeled molecules or of enzymes such as horseradish peroxidase.

discovery by Victor P. Whittaker of the University of Cambridge and Eduardo De Robertis of the University of Buenos Aires that when brain tissue is gently disrupted by being homogenized in a sugar solution, many of the nerve terminals break away from their axons and form intact, closed particles named synaptosomes. The synaptosomes contain the mechanisms of synthesis, storage, release and transmitter inactivation associated with the nerve terminal, and they can be purified from the other neuronal components by spinning them in a centrifuge. This technique has enabled neurochemists to study the mechanisms of synaptic transmission in the test tube.

Perhaps the most far-reaching technical advance has been the development of methods that allow for the selective staining of neurons containing a particular transmitter. One approach is to convert the natural transmitter into a fluorescent derivative that will glow when it is exposed to ultraviolet radiation in the fluorescence microscope. A second approach is to inject radioactively labeled molecules of a transmitter into the brain of an experimental animal, where they are selectively taken up by the nerve terminals that release that transmitter; the radioactive terminals can then be detected by placing thin sections of the tissue on radiation-sensitive film. A third approach exploits the high specificity of antibodies. An enzyme involved in the synthesis of a particular transmitter is purified from brain tissue and injected into an experimental animal, where it induces the manufacture of antibodies that combine with it specifically. The antibodies are then purified, labeled with a fluorescent dye or some other marker and utilized to selectively stain neurons containing the relevant enzyme.

These selective-staining techniques have provided a flood of information about the detailed anatomical distribution of individual transmitters in the complex neuronal circuits of the brain. They have revealed that the transmitters are not distributed diffusely throughout the brain tissue but are highly localized in discrete centers and pathways. The best-mapped transmitters are the monoamines norepinephrine, dopamine and serotonin (so named because each contains a single amine group). As was first shown by Bengt Falck of the University of Lund and Nils-Åke Hillarp of the Karolinska Institute in Sweden, neurons containing monoamines will fluoresce green or yellow if the transmitters are converted into fluorescent derivatives by reacting them with formaldehyde or glyoxylic acid. Such studies have demonstrated that many of the norepinephrine-containing cells in the brain are concentrated in the small cluster of neurons in the brain stem known as the locus coeruleus. The axons of these neurons are highly branched and project to diverse regions, such as the hypothalamus, the cerebellum and the forebrain. The norepinephrine system has been implicated in the maintenance of arousal, in the brain system of reward, in dreaming sleep and in the regulation of mood.

The neurons containing the monoamine transmitter dopamine are concentrated in the regions of the midbrain known as the substantia nigra and ventral tegmentum. Many of the dopamine-containing neurons project their axons to the forebrain, where they are thought to be involved in regulating emotional responses. Other dopamine fibers terminate in the region near the center of the brain called the corpus striatum. In the corpus striatum dopamine appears to play a crucial role in the control of complex movements. The degeneration of the dopamine fibers projecting to this region gives rise to the muscular rigidity and tremors of Parkinson's disease.

The monoamine transmitter serotonin is concentrated in the cluster of neurons in the region of the brain stem known as the raphe nuclei. The neurons of this center project to the hypothalamus, the thalamus and many other brain regions. Serotonin is thought to be involved in temperature regulation, sensory perception and the onset of sleep.

Many other transmitters have been identified, some of which are designated "putative" because their involvement in synaptic transmission in the brain is still somewhat equivocal. For example, several amino acids—the building blocks of proteins—appear to act as transmitters. The common and abundant amino acids glutamic acid and aspartic acid exert powerful excitatory effects on most neurons and may well be the commonest excitatory transmitters at brain synapses. The simplest of all amino acids, glycine, is known to be an inhibitory transmitter in the spinal cord.

The commonest inhibitory transmitter in the brain is gamma-aminobutyric acid (GABA), an amino acid that is not incorporated into proteins. GABA is unique among amino acids in that it is manufactured almost exclusively in the brain and spinal cord. It has been estimated that as many as a third of the synapses in the brain employ GABA as a transmitter. Neurons that contain GABA can be identified in two ways: by labeling them with radioactive GABA or by staining them with antibodies against glutamic acid decarboxylase, the enzyme that catalyzes the manufacture of GABA. It is of interest to note that glutamic acid is a candidate excitatory transmitter in the brain, whereas GABA, which differs from it by a single chemical group, is an inhibitory transmitter. Clearly slight differences in the molecular structure of transmitters can give rise to completely different physiological effects.

The investigation of GABA mechanisms in the brain has been stimulated in recent years by the discovery by Thomas L. Perry of the University of British Columbia that a specific deficit in brain GABA occurs in Huntington's chorea, which is an inherited neurological syndrome. The uncontrolled movements of the disease are caused by progressive deterioration of the corpus striatum in middle life. Postmortem analysis has revealed that the brain damage involves a loss of inhibitory neurons that normally contain GABA, suggesting that a deficit

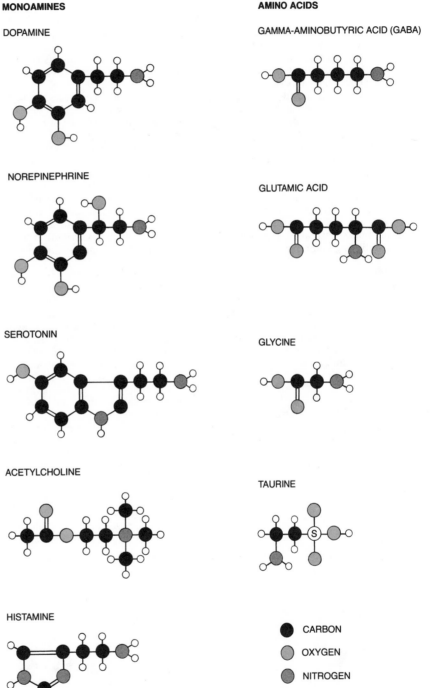

TRANSMITTER CHEMICALS tend to be small molecules that incorporate a positively charged nitrogen atom. Each has a characteristic excitatory or inhibitory effect on neurons, although some transmitters are excitatory in one part of the brain and inhibitory in another. Histamine and taurine are considered putative transmitters because the experimental evidence for them is not yet complete. According to Dale's principle only one transmitter is released from all the terminals of an axon. Exceptions to this principle, however, have been found recently.

of the transmitter might be specifically responsible for the disease. Unfortunately attempts to treat patients by replacing the missing brain GABA is currently not possible, since GABA analogues that are capable of penetrating the blood-brain barrier have not yet been developed.

Recently GABA has also been implicated as a likely target for the actions of antianxiety agents such as diazepam (Valium) and other drugs of the benzodiazepine class. The benzodiazepines are the most widely prescribed of all psychoactive drugs, and their mechanism of action has hitherto not been known. The available evidence suggests that these drugs increase the effectiveness of GABA at its receptor sites in the brain. Although specific diazepam binding sites have been identified in the brain that are clearly distinguishable from the GABA receptors, the two types of receptors appear to interact. The intriguing possibility exists that the brain contains some undiscovered substance that normally acts on the diazepam receptors, perhaps a natural anxiety-producing or -relieving compound.

In addition to identifying the molecular structure and anatomical distribution of the various transmitter substances, neurochemists have made large strides in understanding the precise sequence of biochemical events involved in synaptic transmission. The process of chemical transmission requires a series of steps: transmitter synthesis, storage, release, reaction with receptor and termination of transmitter actions. Each of these steps has been characterized in detail, and drugs have been discovered that selectively enhance or block specific steps. This research has yielded insight into the mechanism of action of psychoactive drugs and also into how certain neurological and mental disorders might be related to specific defects in synaptic mechanisms.

The first step in chemical transmission is the synthesis of the transmitter molecules in the nerve terminals. Each neuron usually possesses only the biochemical machinery it needs to make one kind of transmitter, which it releases from all the terminals of its axon. The transmitter molecules are not manufactured de novo but are prepared by the modification of a precursor molecule, usually an amino acid, through a series of enzymatic reactions.

The manufacture of a transmitter may require one enzyme-catalyzed step (as in the case of acetylcholine) or as many as three steps (for norepinephrine). In the synthesis of norepinephrine the starting material is the amino acid tyrosine, which is taken up into the nerve terminal from the bloodstream. Tyrosine is first converted into the intermediate substance L-DOPA; a second enzyme then converts L-DOPA into dopamine (a transmitter in its own right); a third enzyme converts dopamine into norepinephrine.

After the molecules of the transmitter have been manufactured they are stored in the axon terminal in the tiny membrane-bound sacs called synaptic vesicles. There may be thousands of synaptic vesicles in a single terminal, each of which contains between 10,000 and 100,000 molecules of the transmitter. The vesicles serve to protect the transmitter molecules from enzymes inside the terminal that would otherwise destroy them.

The arrival of a nerve impulse at an axon terminal causes large numbers of transmitter molecules to be discharged from the terminal into the synaptic space. The mechanism of release is still controversial: some investigators believe the synaptic vesicles fuse directly with the presynaptic membrane and discharge their contents into the synaptic space; others contend that a mobile pool of transmitter molecules is liberated through special channels. In any case the nerve impulse is known to trigger release by increasing the permeability of the nerve terminal to calcium ions, which then rush into the terminal and activate the release mechanism.

The released transmitter molecules travel rapidly across the fluid-filled space between the axon terminal and the membrane of the receiving neuron. There they interact with specific receptor sites on the postsynaptic membrane. The receptors are actually large protein molecules embedded in the semifluid matrix of the cell membrane, with parts sticking out above and below the membrane like floating icebergs. A region on the surface of the receptor protein is precisely tailored to match the shape and configuration of the transmitter molecule, so that the latter fits into the former with the precision and specificity of a key entering a lock.

The interaction of the transmitter with its receptor alters the three-dimensional shape of the receptor protein, thereby initiating a sequence of events. The interaction may cause a neuron to become excited or inhibited, a muscle cell to contract or a gland cell to manufacture and secrete a hormone. In each case the receptor translates the message encoded by the molecular structure of the transmitter molecule into a specific physiological response. Some of the responses, such as the contraction of voluntary muscle, take place in a fraction of a second; others, such as the secretion

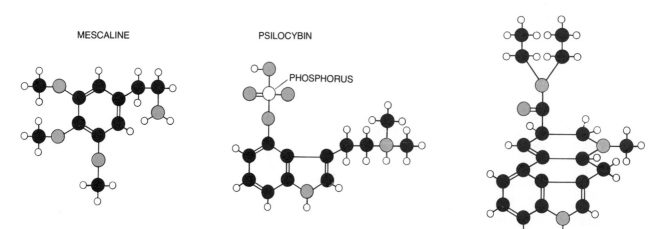

MESCALINE

PSILOCYBIN

PHOSPHORUS

LYSERGIC ACID
DIETHYLAMIDE (LSD)

HALLUCINOGENIC DRUGS show a strong structural resemblance to the monoamine transmitters, suggesting that they may exert their potent effects on consciousness by mimicking these natural transmitters at synaptic receptors in the brain. Mescaline possesses the benzene-ring structure of dopamine and norepinephrine, and psilocybin and LSD incorporate the indole-ring structure of serotonin.

of a hormone, require a span of minutes and sometimes hours.

Many transmitter receptors have two functional components: a binding site for the transmitter molecule and a pore passing through the membrane that is selectively permeable to certain ions. The binding of the transmitter to the receptor changes its shape so that the pore is opened and ions inside and outside the cell membrane flow down their concentration gradients, resulting in either an excitatory or an inhibitory effect on the neuron's firing rate. Whether the electric potential generated by a transmitter is excitatory or inhibitory depends on the specific ions that move and the direction of their movement. Acetylcholine is excitatory at the synapse between nerve and muscle because it causes positively charged sodium ions to move into the cell and depolarize its negative resting voltage. GABA, on the other hand, has a receptor whose pore is selectively permeable to negatively charged chloride ions. When these ions flow through the open pores into the target cell, they increase the voltage across the membrane and temporarily inactivate the cell.

Other transmitters, such as dopamine and norepinephrine, appear to operate by a more elaborate mechanism. In the mid-1950's Earl W. Sutherland, Jr., and his colleagues at Case Western Reserve University demonstrated that these and other transmitters increase or decrease the concentration of a "second messenger" substance in the target cells. The second messenger then mediates the electrical or biochemical effects of the transmitter, or "first messenger." In a discovery that later brought him the 1971 Nobel prize in physiology and medicine, Sutherland identified the second-messenger substance as the small molecule cyclic adenosine monophosphate, or cyclic AMP.

According to Sutherland's hypothesis the receptor protein for norepinephrine (and many other transmitters) is coupled in the target-cell membrane to the enzyme adenylate cyclase, which catalyzes the conversion of the cellular energy-carrying molecule adenosine triphosphate (ATP) into cyclic AMP. Adenylate cyclase is usually inactive, but when norepinephrine binds to the postsynaptic receptor, the enzyme is automatically switched on and begins to rapidly convert ATP into cyclic AMP inside the cell. Cyclic AMP then acts on the biochemical machinery of the cell to initiate the physiological response characteristic of the transmitter.

The second-messenger system is therefore analogous to a relay race, in which the transmitter passes along its message to cyclic AMP at the cell membrane. Of course, the signal is passed not to one but to the many thousands of molecules of cyclic AMP that are gener-

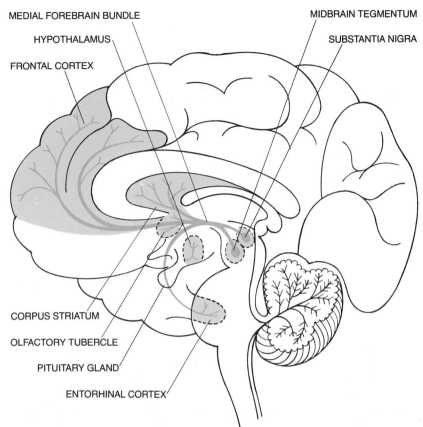

DOPAMINE PATHWAYS in the human brain are shown schematically. The neurons that contain dopamine have their cell bodies clustered in two small regions of the midbrain: the substantia nigra and the tegmentum. These neurons send out widely branching fibers that terminate in the corpus striatum, which regulates motor activity, and in the limbic forebrain, which is involved in emotion. A small set of dopamine neurons in the hypothalamus also regulates secretion of hormones from pituitary. Dopamine has been associated with two brain disorders: a deficiency of the transmitter in the corpus striatum causes the rigidity and tremor of Parkinson's disease, and an excess of dopamine in limbic forebrain may be involved in schizophrenia.

ated by the activated adenylate cyclase associated with each occupied receptor. As a result the very weak signal provided by the transmitter-receptor interaction is amplified several thousandfold inside the cell through the mass production of cyclic AMP.

The application of Sutherland's second-messenger theory to brain function is one of the most exciting areas in neurochemistry today. In 1971 Floyd E. Bloom and his co-workers at the National Institute of Mental Health demonstrated that cyclic AMP could affect signaling in neurons. Later Paul Greengard and his group at the Yale University School of Medicine implicated cyclic AMP in the synaptic actions of several brain transmitters, including norepinephrine, dopamine, serotonin and histamine. Greengard has proposed the unifying hypothesis that cyclic AMP activates specific enzymes in the target cell called protein kinases; these enzymes then act to catalyze the incorporation of phosphate groups into special proteins in the neuronal membrane, altering the permeability of the membrane to ions

and thereby changing the level of excitability of the target cell. Because the second-messenger system works relatively slowly on the time scale of neuronal events it is best suited for mediating the longer-lasting actions of transmitters in the brain such as slow shifts in membrane potential and perhaps the formation of long-term memories. Once cyclic AMP has relayed its message it is inactivated within the cell by the enzyme phosphodiesterase. Drugs that inhibit this enzyme therefore raise the level of cyclic AMP within the target cells and enhance the effect of the transmitter.

To sum up, there appear to be two basic types of transmitter receptor: rapidly acting receptors that mediate the transfer of information by controlling the permeability state of an ion pore, and longer-acting receptors that induce the formation of a second-messenger substance, which in turn mediates the effects of the transmitter inside the target neuron. Many transmitters possess two or more types of receptor. For example, the response to acetylcholine at the synapse between a motor neuron and a muscle cell is mediated by a simple flow of sodium ions across the membrane. In the brain, however, most of the effects of acetylcholine appear to be mediated by another second-messenger molecule: cyclic guanosine monophosphate, or cyclic GMP. Similarly, evidence recently acquired suggests that dopamine acts at two different types of receptor in the brain: the *D*1 receptor, which is coupled to a second-messenger cyclic-AMP system, and the *D*2 receptor, which is not.

Once a transmitter molecule has bound to its receptor it must be rapidly inactivated; otherwise it would act for too long and precise control of

transmission would be lost. Nerve fibers can conduct several hundred impulses per second only if the postsynaptic membrane recovers its resting voltage within a fraction of a millisecond. Some transmitters are inactivated by enzymes situated in the synaptic space. For example, acetylcholine is destroyed by the enzyme acetylcholinesterase, which can cleave 25,000 molecules of the transmitter per second. Norepinephrine is inactivated at the synapse by an entirely different mechanism.

Julius Axelrod and his colleagues at the National Institute of Mental Health found that after norepinephrine is released from the axon terminal it is rapidly pumped back inside. Then the recap-

tured molecules of norepinephrine either are destroyed by the enzymes catechol-*O*-methyltransferase (COMT) and monoamine oxidase (MAO) present in the nerve terminal or are recycled back into the synaptic vesicles. Similar reuptake mechanisms have since been identified for other transmitters, such as dopamine, serotonin and GABA. Reuptake has the obvious advantage over enzymatic degradation in that the transmitter molecules can be conserved through several cycles of release and recapture.

The working out of the steps of synaptic transmission has shed much light on the operation of psychoactive drugs. Some drugs exert their effects by either enhancing or inhibiting the release of a

Met-ENKEPHALIN
(Tyr) (Gly) (Gly) (Phe) (Met)

Leu-ENKEPHALIN
(Tyr) (Gly) (Gly) (Phe) (Leu)

SUBSTANCE P
(Arg) (Pro) (Lys) (Pro) (Gln) (Gln) (Phe) (Phe) (Gly) (Leu) (Met) NH₂

NEUROTENSIN
p(Glu) (Leu) (Tyr) (Glu) (Asn) (Lys) (Pro) (Arg) (Arg) (Pro) (Tyr) (Ile) (Leu)

Ala	ALANINE	Leu	LEUCINE
Arg	ARGININE	Lys	LYSINE
Asn	ASPARAGINE	Met	METHIONINE
Asp	ASPARTIC ACID	Phe	PHENYLALANINE
Cys	CYSTEINE	Pro	PROLINE
Gln	GLUTAMINE	Ser	SERINE
Glu	GLUTAMIC ACID	Thr	THREONINE
Gly	GLYCINE	Trp	TRYPTOPHAN
His	HISTIDINE	Tyr	TYROSINE
Ile	ISOLEUCINE	Val	VALINE

β-ENDORPHIN
(Tyr) (Gly) (Gly) (Phe) (Met) (Thr) (Ser) (Glu) (Lys) (Ser) (Gln) (Thr) (Pro) (Leu) (Val) (Thr) (Leu) (Phe) (Lys) (Asn) (Ala) (Ile) (Val) (Lys) (Asn) (Ala) (His) (Lys) (Lys) (Gly) (Gln)

ACTH (CORTICOTROPIN)
(Ser) (Tyr) (Ser) (Met) (Glu) (His) (Phe) (Arg) (Tyr) (Gly) (Lys) (Pro) (Val) (Gly) (Lys) (Lys) (Arg) (Arg) (Pro) (Val) (Lys) (Val) (Tyr) (Pro) (Asp) (Gly) (Ala) (Glu) (Asp) (Glu) (Leu) (Ala) (Glu) (Ala) (Phe) (Pro) (Leu) (Glu) (Phe) NH₂

ANGIOTENSIN II
(Asp) (Arg) (Val) (Tyr) (Ile) (His) (Pro) (Phe) NH₂

OXYTOCIN
(Ile) (Tyr) (Cys)
(Gln) (Asn) (Cys) (Pro) (Leu) (Gly) NH₂

VASOPRESSIN
(Phe) (Tyr) (Cys)
(Gln) (Asn) (Cys) (Pro) (Arg) (Gly) NH₂

VASOACTIVE INTESTINAL POLYPEPTIDE (VIP)
(His) (Ser) (Asp) (Ala) (Val) (Phe) (Thr) (Asp) (Asn) (Tyr) (Thr) (Arg) (Leu) (Arg) (Lys) (Gln) (Met) (Ala) (Val) (Lys) (Lys) (Tyr) (Leu) (Asn) (Ser) (Ile) (Leu) (Asn) NH₂

SOMATOSTATIN
(Ala) (Gly) (Cys) (Lys) (Asn) (Phe) (Phe) (Trp)
(Cys) (Ser) (Thr) (Phe) (Thr) (Lys)

THYROTROPIN RELEASING HORMONE (TRH)
p(Glu) (His) (Pro) NH₂

LUTEINIZING-HORMONE RELEASING HORMONE (LHRH)
p(Glu) (His) (Trp) (Ser) (Tyr) (Gly) (Leu) (Arg) (Pro) (Gly) NH₂

BOMBESIN
p(Glu) (Gln) (Arg) (Leu) (Gly) (Asn) (Gln) (Trp) (Ala) (Val) (Gly) (His) (Leu) (Met) NH₂

CARNOSINE
(Ala) (His)

CHOLECYSTOKININ-LIKE PEPTIDE
(Asp) (Tyr) (Met) (Gly) (Trp) (Met) (Asp) (Phe) NH₂

NEUROPEPTIDES are short chains of amino acids found in brain tissue. Many of them are localized in axon terminals and are released by a calcium-dependent process, suggesting that they are transmitters. Neuropeptides differ from previously identified transmitters, however, in that they appear to orchestrate complex phenomena such as thirst, memory and sexual behavior. Moreover, they play a multiplicity of roles in different parts of the body. For example, somatostatin inhibits the release of human growth hormone from the pituitary, regulates the secretion of insulin and glucagon by the pancreas and appears to function as a transmitter in the spinal cord and brain.

particular transmitter from axon terminals. For example, the potent stimulant amphetamine triggers the release from nerve terminals in the brain of dopamine, a transmitter associated with the arousal and pleasure systems in the brain. Excessive use of amphetamine by addicts can lead to disruption of thought processes, hallucinations and delusions of persecution, symptoms very similar to those found in some forms of schizophrenia. This and other evidence has led to the hypothesis that an overactivity in the brain dopamine systems may underlie the symptoms of schizophrenia.

Another intriguing finding is that the wide variety of antischizophrenic drugs that have been developed, such as chlorpromazine (Thorazine) and haloperidol (Haldol), share the property of binding tightly to dopamine receptors in the brain, thereby preventing the natural transmitter from activating them. This discovery has proved to be one of the most promising leads in modern schizophrenia research. The latest evidence suggests that schizophrenia is associated with an overproduction of dopamine or an overresponsiveness to the transmitter in certain regions of the brain. Work in my laboratory at the Neurochemical Pharmacology Unit of the British Medical Research Council and by T. J. Crow at the Medical Research Council's Clinical Research Centre in London and by Philip Seeman at the University of Toronto has revealed abnormally high concentrations of dopamine and dopamine receptors in the brains of deceased schizophrenics, particularly in the limbic system, a system of brain regions involved in emotional behavior. The dopamine pathways in these regions may therefore be a primary target for antipsychotic drugs.

Many psychoactive drugs may act by mimicking natural transmitters at their postsynaptic receptors. Many hallucinogenic drugs, for example, bear a structural resemblance to natural transmitters: mescaline is similar to norepinephrine and dopamine, and both LSD and psilocybin are related to serotonin. These drugs may therefore operate on monoamine mechanisms, although their precise modes of action are still not known. LSD is unusual because of its extraordinary potency: as little as 75 micrograms (a barely visible speck) is sufficient to induce hallucinations.

The methylxanthine drugs, such as caffeine and theophylline, are thought to exert their effects by acting through the second-messenger system. Specifically they inhibit the enzyme phosphodiesterase, which degrades cyclic AMP, so that they ultimately increase the amount of cyclic AMP that is generated in response to the transmitter. As a result these drugs exert a general mild stimulant action on the brain. Caffeine is the

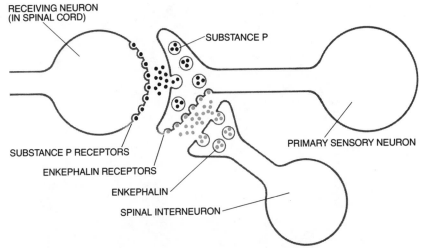

HYPOTHETICAL GATING MECHANISM at the first synaptic relay in the spinal cord may regulate the transmission of pain information from the peripheral pain receptors to the brain. In the dorsal horn of the spinal cord, interneurons containing the peptide transmitter enkephalin make synapses onto the axon terminals of the pain neurons, which utilize substance P as their transmitter. Enkephalin released from the interneurons inhibits the release of substance P, so that the receiving neuron in the spinal cord receives less excitatory stimulation and hence sends fewer pain-related impulses to the brain. Opiate drugs such as morphine appear to bind to unoccupied enkephalin receptors, mimicking the pain-suppressing effects of enkephalin system.

principal active ingredient of coffee and tea; the weaker stimulant theophylline is found primarily in tea. Billions of pounds of coffee and tea are consumed each year, making the methylxanthines among the most widely used drugs.

Finally, certain drugs potentiate the effects of a transmitter by blocking its degradation in the synapse. One such group of drugs is represented by iproniazid (Marsilid) and other drugs that inhibit the enzyme monoamine oxidase, which degrades norepinephrine, dopamine and serotonin. As a result of the blockage of this enzyme the arousing effects of these monoamines are enhanced, accounting for the antidepressant actions of the drugs. A second group of antidepressant drugs, the tricyclics, also amplify the effects of norepinephrine and serotonin in the brain. These drugs, of which the best-known are imipramine (Tofranil) and amitriptyline (Elavil), block the reuptake of norepinephrine and serotonin from the synapse; the stimulant drug cocaine appears to work by the same mechanism. Such observations have suggested that depression may be associated with low levels of amine transmitters at brain synapses, whereas mania may be associated with excessively high levels of these transmitters.

The number of chemical-messenger systems known to exist in the brain has expanded dramatically in recent years with the discovery of a new family of brain chemicals: the neuropeptides. These molecules are chains of amino acids (ranging from two to 39 amino acids

long) that have been localized within neurons and are considered to be putative transmitter substances. Some of them were first identified as hormones secreted by the pituitary gland (ACTH, vasopressin), as local hormones in the gut (gastrin, cholecystokinin) or as hormones secreted by the hypothalamus to control the release of other hormones from the pituitary gland (luteinizing-hormone releasing hormone, somatostatin).

The newest and most exciting of the neuropeptides are the enkephalins and the endorphins: chemicals occurring naturally in the brain that bear a surprising similarity to morphine, the narcotic drug derived from the opium poppy. The discovery of these peptides followed the realization that certain regions of the brain bind opiate drugs with high affinity. The opiate receptors were detected by measuring the binding of radioactively labeled opiate compounds to fragments of neuronal membranes. Three research groups, led by Solomon H. Snyder and Candace B. Pert at the Johns Hopkins University School of Medicine, by Eric J. Simon at the New York University School of Medicine and by Lars Terenius at the University of Uppsala, developed such receptor-labeling techniques and found that opiate receptors were concentrated in those regions of the mammalian brain and spinal cord that are involved in the perception and integration of pain and emotional experience.

Then in 1975 John Hughes and Hans W. Kosterlitz of the University of Aberdeen isolated two naturally occurring

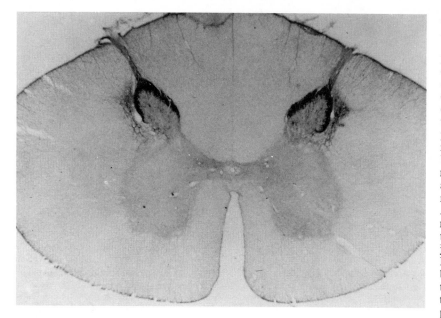

LOCALIZATION OF SUBSTANCE P in the spinal cord of the monkey was achieved by treating the tissue with specific antibodies that were labeled with a dark-staining chemical. The stain is present only in the dorsal horns of the spinal cord, which receive input from the peripheral pain fibers. The micrograph was made by Stephen Hunt of University of Cambridge.

peptides in the brain that bound tightly to the opiate receptors and named them enkephalins. Both enkephalins are chains of five amino acids; they are identical in sequence except for the terminal amino acid, which in one is methionine and in the other is leucine. Other morphinelike peptides, named endorphins, were subsequently isolated from the pituitary gland. Recent experiments have suggested that several procedures employed to treat chronic pain—acupuncture, direct electrical stimulation of the brain and even hypnosis—may act by eliciting the release of enkephalins or endorphins in the brain and spinal cord. This hypothesis is based on the finding that the effectiveness of all these procedures can be largely blocked by the administration of naloxone (Narcan), a drug that specifically blocks the binding of morphine to the opiate receptor.

Many of the neuropeptides found in the mammalian brain have been shown to be concentrated in the terminals of particular sets of neurons, and several are known to be released from axon terminals by a calcium-dependent process. Such findings, together with the observation that minute amounts of neuropeptides exert significant effects on neuronal activity or on the behavior of experimental animals, strongly suggest that these chemicals may well be a new family of transmitters. In most cases, however, the evidence is not yet strong enough to support a definite conclusion.

Perhaps the strongest candidate among the neuropeptides for transmitter status is substance P, a chain of 11 amino acids. It is present in a number of specific neuronal pathways in the brain and also in primary sensory fibers of peripheral nerves. Some of these sensory neurons, whose cell bodies lie in sensory ganglia on each side of the spinal cord, contain substance P and release it from their axon terminals at synapses with spinal-cord neurons. Because substance P excites those spinal neurons that respond most readily to painful stimuli the substance has been suggested to be a sensory transmitter that is specifically associated with the transmission of pain-related information from peripheral pain receptors into the central nervous system.

The morphinelike peptide enkephalin is also present in abundance in small neurons in the part of the spinal cord that receives the fiber input containing substance P. Thomas Jessel and I at the Neurochemical Pharmacology Unit of the Medical Research Council have shown that enkephalin and opiate drugs are able to suppress the release of substance P from sensory fibers. The enkephalin-containing neurons may therefore regulate the input of painful stimuli to the brain by modulating the release of substance P at the first relay in the central nervous system. Similar inhibitory interactions may also take place at higher levels of the brain. Substance P is not the only putative transmitter shown to be localized in sensory neurons; the others so far identified include angiotensin, cholecystokinin, somatostatin and glutamic acid. A bewildering chemical complexity is therefore beginning to emerge as more is learned about the sensory transmitters and their modulating mechanisms in the spinal cord.

A remarkable feature of the neuropeptides in the brain is the global nature of some of their effects. Administration of very small amounts of a neuropeptide (usually directly into the brain to circumvent the blood-brain barrier) can trigger a complex and highly specific pattern of behavior in experimental animals. For example, injection into the brain of nanogram amounts of the neuropeptide angiotensin II elicits intense and prolonged drinking behavior in animals that were not previously thirsty. Another peptide, luteinizing-hormone releasing hormone, induces characteristic female sexual behavior when it is injected into the brain of a female rat. Even more striking, as has been shown by David de Wied and his colleagues at the University of Utrecht, the administration of small amounts of the neuropeptide vasopressin markedly improves the memory of learned tasks in laboratory animals. Preliminary clinical trials of this agent are now in progress to ascertain whether it may have beneficial effects on human patients suffering from memory loss.

It therefore seems that the neuropeptides may be chemical messengers of a character different from that of the previously identified transmitters: they appear to represent a global means of chemical coding for patterns of brain activity associated with particular functions, such as body-fluid balance, sexual behavior and pain or pleasure. An unexpected observation is that biologically active peptides originally found in the gastrointestinal tract, such as gastrin, substance P, vasoactive intestinal polypeptide (VIP) and cholecystokinin, are also present in the central nervous system. Conversely some peptides originally found in the brain have later been found in the gut (somatostatin, neurotensin, enkephalins). It therefore appears that these peptides serve a multiplicity of roles, acting as local hormones or transmitters in the gastrointestinal tract and as global transmitters in the brain. Roger Guillemin of the Salk Institute has suggested that the multiple functions of neuropeptides may be due to the opportunism of the evolutionary process, in which a molecule that serves one function may be adapted to serve another function at a different time and place.

A number of other chemicals appear to play a modulatory role in neuronal communication. The prostaglandins, which consist of a five-carbon ring with two long carbon chains attached to it, are present at high levels in brain tissue and elicit a variety of excitatory and inhibitory effects on neurons, depending on the precise molecular structure of the prostaglandin and the identity of the target cell. Whereas the transmitters have rapid and transient effects, the prostaglandins elicit long-term shifts in the polarization of the neuronal membrane, suggesting that they play a modulatory role rather than a transmitter role. It is possible that they act in concert with transmitters to subtly alter their effects.

Still another set of chemicals plays a nutritive role rather than a messenger role. These "trophic" substances are thought to be secreted from nerve terminals and to maintain the viability of the target cell; other trophic substances are taken up by the nerve terminal and transported along the axon in a retrograde fashion to nourish the neuron itself. The well-known phenomenon of muscle atrophy after an innervating nerve has been severed may result from the inability of the muscle cells to obtain the trophic substances they require. A number of degenerative brain diseases may result from the failure of central neurons to exchange trophic substances. The best-characterized trophic substance to date is the nerve-growth factor (NGF), a protein that is essential for the differentiation and survival of peripheral sensory and sympathetic neurons and that may also be involved in maintaining central monoamine neurons.

Apart from the ever increasing list of chemical transmitters, the variety of different mechanisms by which transmitters can exert their effects is also becoming apparent. For example, instead of directly exciting or inhibiting a target neuron, a transmitter released from one nerve terminal can act presynaptically on an adjacent nerve terminal to increase or decrease the release of transmitter from that nerve terminal. It is also clear that there may be several different types of receptors for a given transmitter substance (some mediated by second-messenger systems and some not), accounting for the diverse excitatory or inhibitory effects of a given transmitter in different parts of the brain. Even the well-established concept (first suggested by Sir Henry Dale) that a neuron releases only one transmitter chemical from all its terminals may not be inviolable: a number of neuropeptides have been found to coexist in the same neurons with norepinephrine or serotonin. The functional significance of such dual-transmitter systems is not yet known. In addition the precise chemical disturbances that underlie such common disorders as epilepsy, senile dementia, alcoholism, schizophrenia and depression remain largely obscure. Although the study of transmitter systems in the brain has already provided major clues to the chemical mechanisms involved in learning, memory, sleep and mood, it seems clear that the most exciting discoveries lie ahead.

4

Neuropeptides

by Floyd E. Bloom
October 1981

They are short chains of amino acids that are active in the nervous system. In some cases they transmit signals between nerve cells and also serve the body as a hormone

The two major systems that coordinate the activity of cells with the needs of the body have long been thought to function in quite different ways. In the nervous system each nerve cell has been taken to affect its set of target cells at synapses: specialized sites at which a chemical messenger—a neurotransmitter—is released by one cell and received by another. In many instances the neurotransmitter is a monoamine: a substance the nerve cell synthesizes by making minor changes in an amino acid. Sometimes the neurotransmitter is the amino acid itself. In any case its release and reception take milliseconds. In short, such communication is point-to-point and fast. In the endocrine system, on the other hand, each gland cell has long been known to release its chemical product—a hormone—into the circulating blood. The hormone is often a peptide: a short chain of amino acids. The release of the hormone, its circulation in the blood and its influence on its target cells throughout the body takes minutes or hours.

These distinctions between the two systems are now in disarray. For one thing norepinephrine, a monoamine neurotransmitter, turns out to be also a hormone: it is released by gland cells in the medulla of the adrenal gland. Conversely, vasopressin, a peptide hormone, turns out to be also a neurotransmitter: nerve cells in the hypothalamus, a part of the brain, rely on vasopressin to signal other nerve cells in the brain. Today more than a dozen cell-to-cell messengers are known to be capable of relaying signals either between nerve cells or between gland cells and their targets. Typically each messenger is discovered first as a factor: a substance of unknown chemical composition that has a physiological effect such as the dilation of arteries or the contraction of muscle. Often it emerges that the factor is made up of amino acids. Then it emerges that the factor is active in the brain. Thus the factor is revealed as a neuropeptide.

The discovery of a peptide messenger begins with the suspicion that a chemical substance is responsible for the interaction of two groups of cells. For example, it has long been known that if an animal is anesthetized with ether, the cortex of the adrenal gland releases steroid hormones in quantity. If the anterior lobe of the pituitary gland is first removed, no such release is measured. Presumably, therefore, a substance released by the anterior lobe causes the adrenal cortex to act. The substance turns out to be adrenocorticotrophic hormone, or ACTH, also called corticotropin. In a different experiment the blood vessels linking the hypothalamus to the anterior lobe of the pituitary are interrupted. Again the etherization of the animal does not give rise to the release of steroids. Evidently the hypothalamus governs the release of pituitary corticotropin. The substance by which it does so is called corticotropin releasing factor. Its molecular structure has not yet been worked out.

The Process of Discovery

The suspicion that two groups of cells interact by means of a chemical messenger motivates a sequence of experiments that is well established today. The sequence was devised largely by Vincent du Vigneaud of the Cornell University Medical College as he worked in the 1940's and 1950's to identify the hormones secreted by the posterior lobe of the pituitary. It was elaborated in the 1960's and 1970's by Roger Guillemin, who is now at the Salk Institute for Biological Studies, and by Andrew V. Schally, who is now at the New Orleans Veterans Administration Hospital. The objects of their search were the hormones by which the hypothalamus regulates the anterior lobe of the pituitary.

First an extract is prepared from the group of cells that presumably release a factor. This can be done, for example, by homogenizing the cells. The extract is applied to the tissue whose cells the factor controls. The potency of the extract is noted; then the extract can be refined by passing it through a chemical sieve consisting of a gel that selectively filters molecules on the basis of their size or their electric charge. The refined extract is applied to the tissue. With skill, and often with luck, one finds that it now has greater potency than the original extract had. The factor has therefore been concentrated. Further refinements of the extract are tested for potency. Ultimately the factor is purified. In some instances the process begins with extracts from hundreds of thousands of fragments of brain and ends with a few nanograms of an unknown chemical substance.

At this point a different technology is applied to the factor to determine its chemical structure. Suppose the factor is inferred to be a peptide because it loses its biological activity when it is treated with enzymes that cleave peptide chains at the linkages between two successive amino acids. A chemical assay then shows the proportions of the various amino acids. The precise sequence of amino acids remains to be determined. The classical technique is to treat the factor with a number of enzymes each of which cleaves the peptide chain at the linkages between two specific amino acids.

The resulting fragments are collected. Individual amino acids are successively cleaved from each one and identified by properties such as the rate at which they travel in a column of gel that filters molecules according to their electric charge. In this way the sequence of amino acids in each fragment is worked out. Next the order of the fragments themselves is determined by finding overlapping sequences of amino acids in the sets of fragments cleaved from the factor by enzymes that cleave a peptide at different sites. The newest technique removes amino acids from the full peptide chain one by one; a mass spectrometer then identifies each amino acid in the sequence by its weight.

When a sequence has been deter-

PRESENCE OF ENKEPHALIN in the organ of sight of several species is demonstrated by a technique that tags the enkephalin with an antibody and then tags the antibody with a second antibody to which a fluorescent molecule is bound. The enkephalin itself is a neuropeptide consisting of five amino acids. The upper photograph shows a retinula, or sight organ, of the lobster. The fluorescence marks sprays of cells. Each spray is an ommatidium, a cluster of light receptors some 10 micrometers in diameter. The photograph was made by Jorge Mancillas and Jacqueline F. McGinty of the Salk Institute. The lower photograph shows a cross section of the retina of a chick. Here the cells containing enkephalin are nerve cells, not light receptors. Each cell belongs to the class of neurons called amacrine cells, which make local connections in the retina by means of filaments that the technique also renders fluorescent. Some of the amacrine cells not marked by fluorescence in this preparation turn out to contain other neuropeptides: neurotensin, substance P, somatostatin or vasoactive intestinal peptide. The photograph was made by Nicholas Brecha and Harvey J. Karten at the State University of New York at Stony Brook.

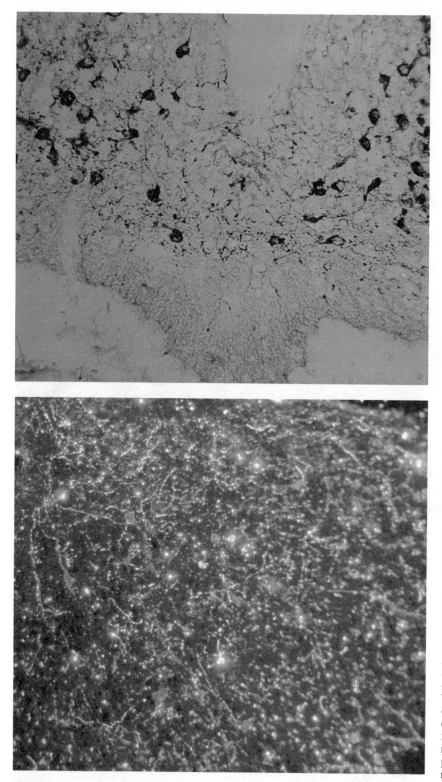

PRESENCE OF BETA-ENDORPHIN, a neuropeptide consisting of a chain of 31 amino acids, is demonstrated in the cell group in the rat hypothalamus called the arcuate nucleus by a technique that employs antibodies to tag the cells containing beta-endorphin with the enzyme horseradish peroxidase. The enzyme catalyzes the polymerization of the brown pigment diaminobenzidine. The upper photograph shows the preparation under the light microscope. The brown pigment marks numerous nerve cells. They appear to be the only cells in the brain of the rat that contain beta-endorphin. Each cell is some 25 micrometers across. The space at the top of the field is the third ventricle, a fluid-filled chamber inside the brain. The lower photograph shows the same tissue under dark-field illumination. Light passes obliquely through the tissue and is deflected onto the photographic film by the particles of the pigment. The varicosities that show up in gold are axons and dendrites: the filamentous extensions of the cells marked by the staining technique. The cell bodies themselves are small, elongated triangles. Bluish pinpoints of light are irregularities at the tissue's surface. Some of them are capillaries cut in cross section. The photographs were made by Elena Battenberg and the author at the Salk Institute.

mined, its proportional content of the various amino acids is matched against the composition of the factor as it has been determined by chemical assay. Then the peptide is synthesized. Today this can be done readily; indeed, a peptide is often commercially available within weeks of its discovery. The quantity of the synthetic replicate is much greater than the quantity investigators could hope to produce by progressive purifications of a cell extract. Hence the replicate can be tested to see if it matches the purified natural peptide in both action and potency. Moreover, the replicate can be tested for biological activity on tissues other than the one it is known to act on.

Further still, the replicate can be injected into an animal of a species other than the one from which the peptide was purified. The immune system of the animal will prime antibodies against the peptide. The peptide itself can be quite short; thus its sequence of amino acids may be nearly identical in many species. In such instances the priming of antibodies is encouraged by linking the peptide to a large carrier molecule before it is injected. In any case the antibodies have several uses. In the technique called the radio-immunoassay, developed by Solomon A. Berson and Rosalyn S. Yalow of the Bronx Veterans Administration Hospital, they are mixed with a replicate of the peptide in which some of the atoms are radioactive. The antibodies bind to the replicate. Then a tissue extract is added to the mixture. The native peptide in the extract will displace a certain amount of the replicate. The degree of displaced radioactivity is a sensitive measure of the amount of native peptide.

The antibodies can also serve as a microscopic stain to reveal the location of cells that store the peptide and presumably utilize it. In some techniques the antibody is labeled with a chemical group that is fluorescent or one that is radioactive. In still other techniques the antibody is linked to an enzyme that can manufacture a pigment inside cells. The most elaborate staining technique calls for animals of three different species. Suppose a peptide has been purified from the brain of a rat and a synthetic replicate of the peptide has been prepared. The replicate is injected into a rabbit. The resulting rabbit antibodies are applied to sections of the brain of a rat. There they bind to the native peptide. The rabbit antibodies are also injected into a goat. The result is goat antibodies that act against rabbit antibodies. These goat molecules are labeled and applied to the sectioned rat brain. They bind to the rabbit antibody, which is already bound to the peptide. The double-antibody technique has the advantage that the antibody prepared against the synthetic peptide is not subjected to a chemical labeling reaction, which might

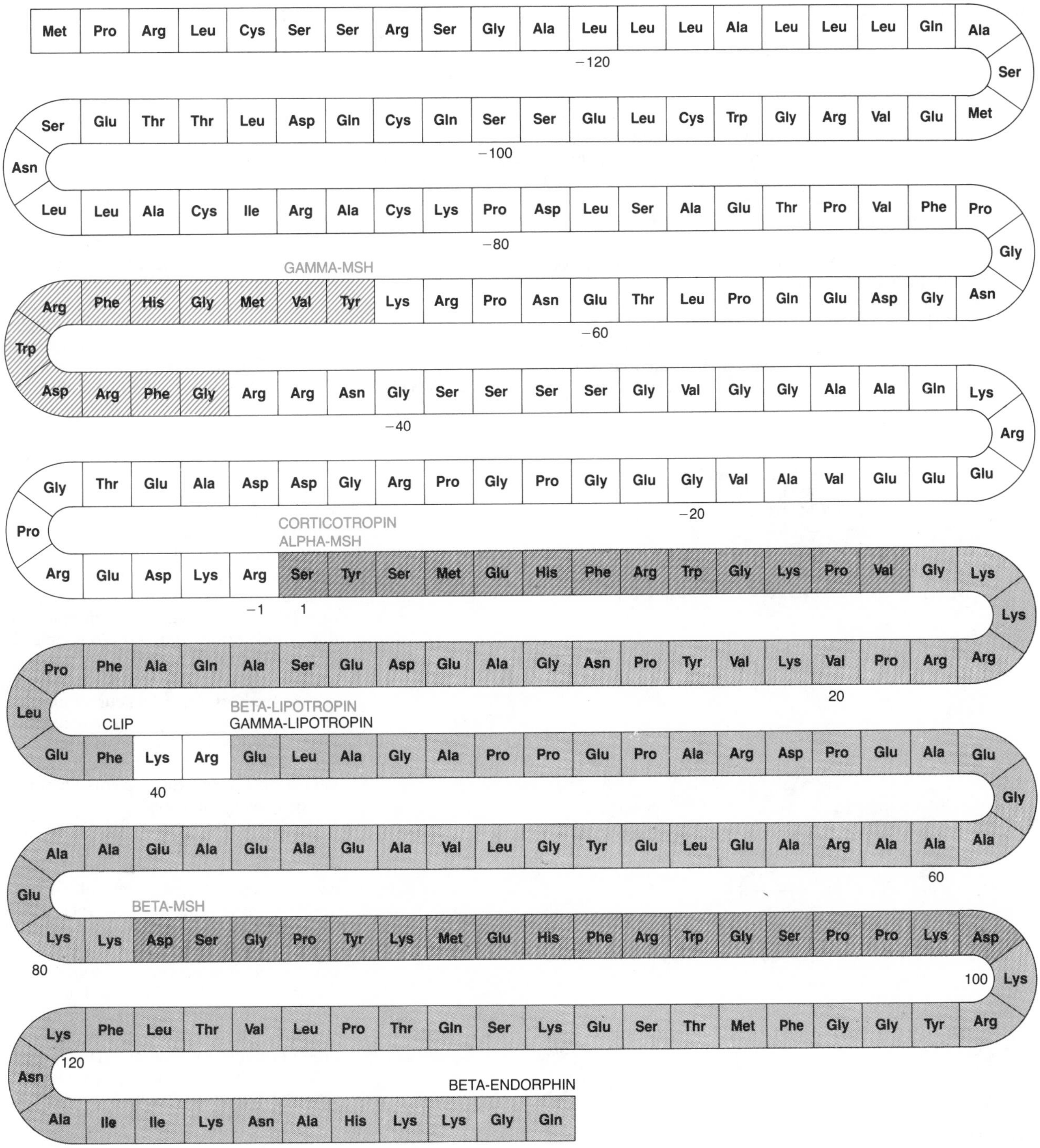

Met | Pro | Arg | Leu | Cys | Ser | Ser | Arg | Ser | Gly | Ala | Leu | Leu | Leu | Ala | Leu | Leu | Leu | Gln | Ala — Ser
−120

Ser — Glu | Thr | Thr | Leu | Asp | Gln | Cys | Gln | Ser | Ser | Glu | Leu | Cys | Trp | Gly | Arg | Val | Glu | Met
−100
Asn

Leu | Leu | Ala | Cys | Ile | Arg | Ala | Cys | Lys | Pro | Asp | Leu | Ser | Ala | Glu | Thr | Pro | Val | Phe | Pro — Gly
−80
GAMMA-MSH

Arg | Phe | His | Gly | Met | Val | Tyr | Lys | Arg | Pro | Asn | Glu | Thr | Leu | Pro | Gln | Glu | Asp | Gly — Asn
Trp — Asp −60

Asp | Arg | Phe | Gly | Arg | Arg | Asn | Gly | Ser | Ser | Ser | Ser | Gly | Val | Gly | Gly | Ala | Ala | Gln — Lys
−40 — Arg

Gly | Thr | Glu | Ala | Asp | Asp | Gly | Arg | Pro | Gly | Pro | Gly | Glu | Gly | Val | Ala | Val | Glu | Glu — Glu
Pro −20
CORTICOTROPIN / ALPHA-MSH

Arg | Glu | Asp | Lys | Arg | Ser | Tyr | Ser | Met | Glu | His | Phe | Arg | Trp | Gly | Lys | Pro | Val | Gly — Lys
−1   1 — Lys

Pro | Phe | Ala | Gln | Ala | Ser | Glu | Asp | Glu | Ala | Gly | Asn | Pro | Tyr | Val | Lys | Val | Pro | Arg — Arg
Leu 20
CLIP / BETA-LIPOTROPIN / GAMMA-LIPOTROPIN

Glu | Phe | Lys | Arg | Glu | Leu | Ala | Gly | Ala | Pro | Pro | Glu | Pro | Ala | Arg | Asp | Pro | Glu | Ala — Glu
40 — Gly

Ala | Ala | Glu | Ala | Glu | Ala | Glu | Ala | Val | Leu | Gly | Tyr | Glu | Leu | Glu | Ala | Arg | Ala | Ala — Ala
Glu — 60
BETA-MSH

Lys | Lys | Asp | Ser | Gly | Pro | Tyr | Lys | Met | Glu | His | Phe | Arg | Trp | Gly | Ser | Pro | Pro | Lys — Asp
80 — 100 — Lys

Lys | Phe | Leu | Thr | Val | Leu | Pro | Thr | Gln | Ser | Lys | Glu | Ser | Thr | Met | Phe | Gly | Gly | Tyr — Arg
Asn — 120
BETA-ENDORPHIN

Ala | Ile | Ile | Lys | Asn | Ala | His | Lys | Lys | Gly | Gln

Ala Alanine
Arg Arginine
Asn Asparagine
Asp Aspartate
Cys Cysteine
Gln Glutamine
Glu Glutamate
Gly Glycine
His Histidine
Ile Isoleucine
Leu Leucine
Lys Lysine
Met Methionine
Phe Phenylalanine
Pro Proline
Ser Serine
Thr Threonine
Trp Tryptophan
Tyr Tyrosine
Val Valine

SEVERAL PEPTIDE MESSENGERS lie in the single protein (a long peptide chain) whose amino acid sequence in cattle is diagrammed. The protein, pro-opiomelanocortin, is synthesized by cells in the arcuate nucleus of the hypothalamus and also by cells in an endocrine gland: the anterior lobe of the pituitary. Evidently it serves as a precursor from which any of several intercellular messenger substances can be cleaved. Specifically the pro-opiomelanocortin chain incorporates the chain of corticotropin (1–39), a hormone released by the anterior lobe that stimulates the cortex of the adrenal gland to release other hormones, and the chain of beta-lipotropin (42–132), a weak hormone released by the anterior lobe that causes fat cells to break down their lipids. The corticotropin chain in turn incorporates the chain of alpha-melanocyte-stimulating hormone, or alpha-MSH (1–13), which in the frog is known to change the color of the skin. It also incorporates the chain of CLIP, or corticotropin-like

intermediate-lobe peptide (18–39), a weak form of corticotropin. At the same time beta-lipotropin incorporates gamma-lipotropin (42–99), a weak form of beta-lipotropin. It also incorporates a version of MSH: beta-MSH (82–99). Further still, it incorporates endorphins. Beta-endorphin (102–132) is thought to be a neurotransmitter: a substance by which a nerve cell signals other cells. Five amino acids at one end of the endorphin chain (102–106) duplicate the sequence of an enkephalin. A third version of MSH (gamma-MSH) lies in a part of pro-opiomelanocortin not identified as a messenger. The ends of gamma-MSH may lie at a bond between the amino acids arginine and lysine (−57 and −56) and a bond between two arginines (−43 and −42), places where enzymes that cleave peptide chains typically act. The amino acid sequence in pro-opiomelanocortin was deduced by Stanley N. Cohen of the Stanford University School of Medicine and Shosaku Numa and Shigetada Nakanishi of Kyoto University.

alter its ability to react with the native peptide molecule.

The employment of antibodies to seek out the peptide often reveals that many more cells contain the peptide than had been thought. Indeed, certain peptides have been found in neurons that were known to contain a monoamine and so had been taken to be specialized for the release of that neurotransmitter alone. At the same time the elucidation of the chemical structure of the factor often makes possible an improvement in the method used to purify the natural factor from the cells that contain it. If the factor is a peptide, for example, and the original method of purification began with a homogenate of brain tissue, it may well be that the disruption of the

cells by the homogenization allowed the peptide to be attacked by the enzymes called peptidases. A method can then be devised in which such enzymes are inactivated before the peptide is extracted. Sometimes the result is surprising: a larger form of the peptide is discovered. It had gone undetected because the first extraction procedure had destroyed it. In some instances the larger peptide is more potent than the smaller one.

Vasopressin and Oxytocin

As more and more peptides have been purified and their structure has been determined several groups of chemically related substances have emerged. Two types of group can be distinguished. In

one type the peptides purified from cells across a wide range of species include almost identical long sequences of amino acids. In the second type a number of different large peptides purified from the cells of a single species include identical short sequences of amino acids. Presumably the sequences that are identical have proved particularly well suited for intercellular signaling through long periods of evolution.

Consider the peptide messengers first identified by du Vigneaud and his colleagues as factors released by the posterior lobe of the pituitary gland, an appendage to the brain. They can be seen in the gland as fatty droplets inside the terminals of the axons, or long fibers, of nerve cells in the hypothalamus. To Ernst A. Scharrer, who was then at the University of Colorado School of Medicine, the cells seemed to be typical neurons: they appeared to get information from their synapses with other cells. Scharrer thus advanced the surprising hypothesis that the cells were neurosecretory neurons: like any other nerve cell they were controlled by synaptic connections, and yet on suitable command they could secrete a hormone into the bloodstream. It is now recognized that their axons also project to several levels of the brain. Hence they evidently rely on the hormone as a neurotransmitter. Working at the Albert Einstein College of Medicine in New York, Berta V. Scharrer, Ernst's widow, found similar neurons in the nervous system of a number of invertebrate species.

Du Vigneaud named the factors in accord with their physiological action. Specifically, he distinguished a factor he called antidiuretic hormone, which acts on the kidneys of an animal deprived of drinking water to lessen the loss of water in the urine, and a factor called oxytocin, which promotes the contraction of the muscle of the uterus and thereby speeds birth. When it emerged that antidiuretic hormone could elevate the blood pressure by causing certain arteries to constrict, the factor was given the second name vasopressin.

Vasopressin and oxytocin turn out, then, to be an example of a family of peptides whose structure is well conserved across a wide range of species. In most vertebrate animals less advanced than the mammals the family is represented by a single substance, which is called arginine-vasotocin because of the presence of the amino acid arginine at position No. 8 in its chain of amino acids. In mammals the hormones oxytocin and vasopressin appear. Oxytocin is identical with vasotocin except that the amino acid leucine takes the place of arginine at position No. 8. Vasopressin is slightly more idiosyncratic. In all mammals except the pig it is identical with vasotocin except that phenylalanine substitutes for isoleucine at position No. 3. In the pig one finds two sub-

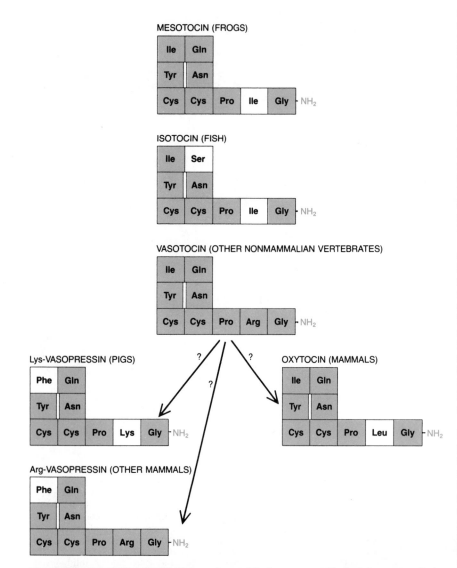

VASOPRESSIN AND OXYTOCIN are the peptide hormones synthesized by nerve cells in the hypothalamus of mammals and then released by those cells from the ends of their axons in the posterior lobe of the pituitary gland. Vasopressin (*lower left*) constricts arteries and causes the kidneys to recapture water from the urine. Oxytocin (*lower right*) differs only slightly in structure from vasopressin, yet its action is quite different: it promotes contractions of the uterus and thereby speeds birth. Both vasopressin and oxytocin are almost identical in structure with vasotocin, mesotocin or isotocin (*top*), the hormones released from the posterior lobe of the pituitary gland in nonmammalian vertebrate animals. The similarities (*color*) suggest that vasopressin and oxytocin arose through mutations in duplicate copies of a single gene.

stitutions: lysine takes the place of arginine at position No. 8 and phenylalanine takes the place of isoleucine at position No. 3.

Wilbur H. Sawyer of the Columbia University College of Physicians and Surgeons has conjectured that a gene duplication may have allowed a peptide modified at position No. 8 to evolve from arginine-vasotocin. The modification would lessen the peptide's antidiuretic action and enhance its oxytocic action. In any case the alteration in the action of a peptide hormone because of a minor change in the structure of the peptide shows that each peptide acts on different target cells in spite of the small degree of difference. It follows that the receptors on the target cells (that is, the sites on the membrane of the target cells that "recognize" the messenger) must be precise enough to accept the one molecule and reject the other. Perhaps some further minor modifications of the molecules will yield synthetic peptides whose specificity of action is greatly enhanced. Such substances might well be valuable pharmaceuticals.

Endorphins and Enkephalins

Perhaps the most striking example of the second type of family relationship among peptide messengers—the presence of certain short sequences of amino acids common to several coexistent peptides—is provided by the endorphins and the enkephalins. The terms have emerged from a decade of almost frenzied research; they are now taken to signify classes of brain peptides whose effects on cells resemble those of opiates such as morphine. (The word endorphin is a contraction of endogenous morphine.) The saga of the endorphins and the enkephalins began with the discovery in several laboratories, including those of Solomon H. Snyder at the Johns Hopkins University Medical School, Eric J. Simon at the New York University School of Medicine and Lars Terenius at the University of Uppsala, that certain cells in the brain have receptors that bind opiates. Some of the cells lie in brain structures implicated in the perception of pain; others, however, do not. Then Hans W. Kosterlitz and John R. Hughes of the University of Aberdeen purified from brain-cell extracts a fraction that occurs naturally in the brain and suppresses the contraction of muscle tissue from the intestine, much like morphine itself. The molecules with this action turned out to be two pentapeptides. One of them has the sequence tyrosine-glycine-glycine-phenylalanine-methionine; it is known as met-enkephalin. The other, which is identical except that leucine substitutes for methionine, is leu-enkephalin.

Soon the met-enkephalin sequence was found in a number of longer peptides purified from an endocrine gland: the anterior lobe of the pituitary. Then the longer peptides—the endorphins—

PEPTIDES RESEMBLING GLUCAGON form a family of four. Glucagon (a) is a hormone synthesized by the alpha cells of the islets of the pancreas. It stimulates the liver to break down glycogen into glucose; hence it counteracts insulin. Gastric inhibitory peptide (b) has been purified from the lining of the stomach. It inhibits the contraction of smooth muscle in the small intestine, perhaps to allow time for intestinal enzymes to act on food. Secretin (c) has been purified from the lining of the small intestine. It causes the acinar cells of the pancreas to secrete fluid that neutralizes the acidity of gastric secretions. Vasoactive intestinal peptide, VIP (d), was first purified from the lining of the small intestine. Evidently it dilates the blood vessels there; thus it increases the flow of blood in the intestinal wall. Recently VIP has turned up in nerve fibers that innervate the intestine. It has also been found in the brain. The amino acid sequences shown in the illustration are those of peptides purified from the pig. Duplications of amino acids in the glucagon family are shown in color.

were found also in nerve cells. Some investigators surmised that such nerve cells shorten an endorphin to make an enkephalin. The hypothesis was contradicted, however, by the preparation of antibodies to the endorphins and the enkephalins and the subsequent mapping of the nerve cells that contain them. The staining patterns reported for the enkephalins by Tomas G. M. Hökfelt and his associates at the Karolinska Institute in Sweden show that cells containing enkephalins are widespread in the brain. They lie, for example, in the spinal cord, the brain stem, the hippocampus and the corpus striatum. In contrast the staining patterns my colleagues and I at the Salk Institute have found for the endorphins place them exclusively in nerve cells at the base of the hypothalamus and in endocrine cells in the anterior lobe of the pituitary.

The family relations being discovered within the endorphins and within the enkephalins are becoming complex and distinctive. For the endorphins the first hint of complexity came when Richard Mains and Betty Eipper of the University of Colorado and Nicholas Ling of the Salk Institute discovered why endorphins are found in the cells of the anterior pituitary that release the hormone corticotropin. Each such cell makes a long precursor peptide from which a molecule of endorphin and a molecule of corticotropin are cleaved. The precursor was given the provisional name pro-opiocortin. Yet it was twice as long as it would be if it consisted of a molecule of corticotropin joined to a molecule of endorphin.

Quite recently the entire sequence of amino acids that make up the precursor has been determined by Stanley N. Cohen of the Stanford University School of Medicine and Shosaku Numa and Shigetada Nakanishi of the Kyoto University Faculty of Medicine. These workers determined the sequence without attempting to purify the peptide as in the orthodox process of discovery. Instead they employed the new techniques of genetic engineering to produce multiple copies of the messenger RNA that specifies the sequence. Then they worked out the structure of the messenger RNA. The RNA is a strand of the units called nucleotides. A particular triplet of nucleotides encodes the identity of a particular amino acid.

In the wake of the work of this group it proved advisable to call the precursor pro-opiomelanocortin. The added term melano refers to a sequence of seven amino acids. One copy of the sequence is included in corticotropin; it is called alpha-melanocyte stimulating hormone (alpha-MSH) because it is known to disperse the pigment melanin in the pigment cells (melanocytes) in the skin of the frog. It thereby modulates the color of the skin. (When a green frog is placed in the dark, its skin turns brown.) A nearly identical sequence lies next to the endorphin sequence in pro-opiomelanocortin; it is called beta-MSH. It was discovered by Choh Hao Li of the University of California at San Francisco, who purified a number of pituitary peptides and determined their sequences of amino acids by classical methods in the late 1950's and early 1960's.

What Cohen, Numa and Nakanishi discovered was a third sequence almost identical with the other two. It lies in the section of the pro-opiomelanocortin chain whose sequence of amino acids had not previously been determined. The third sequence was named gamma-MSH in advance of any evidence that it has its own action in the body. Work done at the Salk Institute by Ling and Guillemin and their colleagues and by my colleagues and me suggests, however, that gamma-MSH is stored in cells of the anterior lobe of the pituitary. Moreover, when gamma-MSH is injected into a cerebral ventricle of an experimental animal (a fluid-filled space in the brain), the animal's body temperature decreases. Evidence is also accumulating that certain cells of the hypothalamus rely on gamma-MSH as a neurotransmitter.

Molecules Incorporating Enkephalin

The complexities in the enkephalin branch of the family are also increasing now that investigators in the U.S. and Japan have begun to detect enkephalin sequences in longer molecules that are otherwise unlike the endorphins. The differences lie in the amino acids next to the end of the enkephalin pentapeptide designated the C terminus. Specifically, the endorphins in pro-opiomelanocortin have serine and threonine in that position, whereas the recently discovered molecules that incorporate the enkephalin sequence have lysine and arginine there. Lysine and arginine are special amino acids in that they each have two amino (NH_2) groups, one of which projects from the amino acid's molecular structure. Studies of a large number of peptides show that the sites where two such amino acids are adjacent in the peptide chain are the places where enzymes most often cleave the peptide. The two adjacent amino groups protruding from the peptide chain may well be the feature the enzyme recognizes when it binds to the peptide and cleaves it. The occurrence of the feature next to the enkephalin pentapeptide suggests that the large molecules are precursors of enkephalin.

One version of a large molecule that includes leu-enkephalin was discovered recently by Avram S. Goldstein and his colleagues at the Stanford School of Medicine. They call it dynorphin because it is more potent than enkephalin in the standard biological assay for the effect of an opiate. (In the standard as-

a PHYSAELEMIN

pGlu	Ala	Asp	Pro	Asn	Lys	Phe	Tyr	Gly	Leu	Met

b ELEIDOSIN

pGlu	Pro	Ser	Lys	Asp	Ala	Phe	Ile	Gly	Leu	Met

c SUBSTANCE *P*

Arg	Pro	Lys	Pro	Gln	Gln	Phe	Phe	Gly	Leu	Met

d BOMBESIN

pGlu	Gln	Arg	Leu	Gly	Asn	Gln	Trp	Ala	Val	Gly	His	Leu	Met

e NEUROTENSIN

pGlu	Leu	Tyr	Glu	Asn	Lys	Pro	Arg	Arg	Pro	Tyr	Ile	Leu

PEPTIDES RESEMBLING PHYSAELEMIN form a family of perhaps as many as five. Physaelemin (*a*) has been purified from the skin of the frog. Eleidosin (*b*) has been purified from the salivary glands of the octopus. The other three peptides—substance *P* (*c*), bombesin (*d*) and neurotensin (*e*)—have been purified from the nervous system of mammals. Although all five substances promote the contraction of muscle tissue taken from the viscera, it is not known whether they act in that way in the body. Again the sequences are those of peptides in the pig, and color denotes duplications of amino acids. The abbreviation pGlu stands for pyroglutamate, a form of glutamate in which a side chain on the amino acid binds to one end of the amino acid to form a small ring. Pyroglutamate appears at the same end of four of the peptides.

say the substance is applied to a sample of smooth muscle from the intestine of a guinea pig. The muscle is induced to contract by electrical stimulation, and the degree of the substance's ability to suppress the contraction constitutes the assay.) Other large molecules that include enkephalins are also reported to be more potent than the enkephalin sequence itself. Thus it is conceivable that the first estimates of the potency of the enkephalins were based on studies of the partially degraded larger forms.

Still another complication in the enkephalins arose when Hökfelt and his colleagues applied the immunohistochemical stains for the enkephalins to the cells of the adrenal medulla. The cells were previously known to secrete only the monoamines epinephrine and norepinephrine. Now Hökfelt's group found that these cells show a large degree of what Hökfelt was careful to call "enkephalin-like" reactivity to the stain. He was cautious in his conclusions because he considered it possible that the antibody employed for the staining procedure had bound itself not to enkephalin but to a similar sequence in some undiscovered peptide.

The caution was well placed. A series of discoveries made in laboratories including that of Sidney Udenfriend at the Roche Institute of Molecular Biology has now revealed that several large peptides from extracts of the adrenal medulla include the enkephalin sequence. One of them is at least 10 times larger than enkephalin. It has several copies of the met-enkephalin sequence and at least one copy of leu-enkephalin. The large polyenkephalin peptides seem to be secreted in tandem with epinephrine and norepinephrine but at less than a hundredth the concentration. Moreover, no interaction between the peptide messengers and the monoamine messengers on a common target cell has yet been discovered.

Gut Peptides in the Brain

The nerve cells that include the proopiomelanocortin peptide present a special problem: Do they pare that precursor down into an endorphin, into corticotropin or into gamma-MSH, or do they pare it down into other peptide molecules not yet known to be messengers? Perhaps the signals arriving at such a cell can direct the cleaving of the precursor so that certain axon terminals release specific products.

A related problem then arises. The enkephalins and some of the endorphins suppress the perception of pain. Hence it has sometimes been advanced that the modulation of the perception of pain is their primary function. The enkephalins and the endorphins are now found, however, in brain circuits that are implicated in a wide range of functions, including the control of blood pressure and body temperature, the regulation of the secretion of hormones and the governance of body movement. This wide range of functions has confounded the effort to attribute to the messengers a single domain of function.

To be sure, it might be advanced that several independent systems depend on messengers that share the amino acid sequence of an enkephalin. Presumably the messengers share the sequence because they evolved from some single messenger early in animal evolution. Today each such system would act on a different set of target cells; thus each set of target cells would require its own type of receptor, and a given type of receptor would have to be able to distinguish between two molecules that differ only slightly in structure.

The view that the various systems exploiting enkephalins and endorphins might be independent was presaged by William D. Martin of the University of Kentucky Medical Center well before the endorphins were discovered. His study of the actions of opiates on the spinal cord of the dog had shown that different actions of morphine are simulated or blocked by different drugs. This suggested distinct classes of opiate receptors. More recently the work of Kosterlitz at Aberdeen, Albert Herz at the Max Planck Institute for Psychiatry in Munich, Goldstein at Stanford and several other investigators shows that the various peptides containing an enkephalin pentapeptide differ considerably in their potency in suppressing the perception of pain or the contractility of smooth muscle. This too suggests several classes of receptors.

Adopting (at least for the moment) the view that the cells secreting structurally similar peptides need not be functionally related may help to explain why many peptides once thought to serve only the gut or the endocrine system are also detected in the brain. Often they are found in parts of the brain that have nothing to do with the realm in which the substance acts in the periphery of the body. Among these substances are the peptides of the gut (and now the brain) called substance P, vasoactive intestinal peptide and cholecystokinin. Each one has been found in neurons that make local synaptic connections ("local-circuit neurons") in the cerebral cortex and

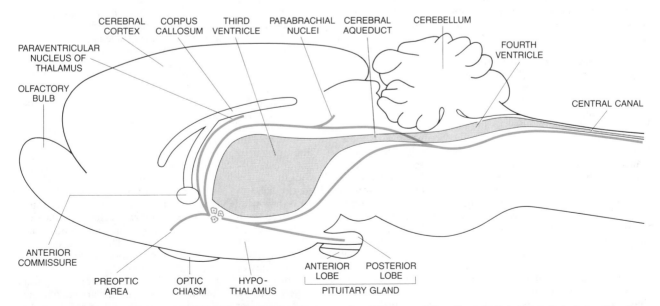

NERVE CELLS EMPLOYING OXYTOCIN (evidently) as their neurotransmitter are found in the rat in the cell groups of the hypothalamus called the supraoptic nucleus and the paraventricular nucleus. The axons of the cells project into the posterior lobe of the pituitary gland, from which they release the oxytocin as a hormone. They also project, however, into cell groups in the brain and the spinal cord.

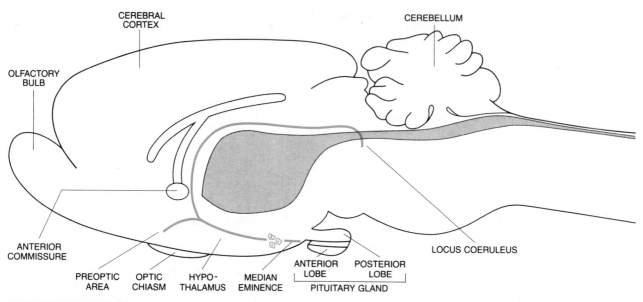

NERVE CELLS CONTAINING ENDORPHIN are found in the rat in the arcuate nucleus of the hypothalamus. The axons of the cells (*also shown in the photographs on page 116*) **project along pathways in the brain similar to those of the cells containing oxytocin. The pattern shown emerges also for corticotropin and gamma-MSH, two other peptide messengers in the precursor pro-opiomelanocortin.**

in the hippocampus. Moreover, nerve fibers containing substance *P* or vasoactive intestinal peptide have been found innervating visceral tissues such as parts of the lung. They have even been found innervating the thyroid gland, where no nerve fibers were known to terminate.

A still more striking example of an almost ubiquitous peptide messenger is somatostatin. It is a substance that Paul Brazeau, Wyley Vale, Roger C. Burgus and Ling and Guillemin of the Salk Institute discovered by applying the classical process of discovery to extracts from the hypothalamus. The biological assay they resorted to as a criterion for successive purification of the extract was the ability of each successive extract to suppress the secretion of growth hormone by cells these workers had cultured from the anterior lobe of the pituitary. Soon it was determined that somatostatin is a peptide consisting of 14 amino acids. Then a replicate of somatostatin was synthesized and antibodies to it were developed. The employment of the antibodies as a stain placed somatostatin in certain cells of the hypothalamus positioned so that they can secrete the substance into a system of capillaries that carries it into the anterior lobe.

Further studies of the distribution of somatostatin have been little less than astounding. In the nervous system somatostatin is found in nerve cells in almost every neural tissue from the cerebral cortex to the autonomic ganglia: the nests of nerve cells in the periphery that govern the tissues of the viscera. In the gut somatostatin is found in the cells that line the intestine. In the endocrine system somatostatin is found in the delta cells of the islets of Langerhans. Its action in the pancreas appears to sup-

press the secretion of the hormones insulin and glucagon. Quite recently larger forms of somatostatin have been extracted from the pancreas and the brain that are more potent than the sequence of 14 amino acids.

Peptides as Neurotransmitters

The utilization of a peptide by a nerve cell as a neurotransmitter differs from the utilization of an amino acid or a monoamine in a fundamental respect: the way the nerve cell synthesizes the substance and conserves it. A neurotransmitter such as gamma-aminobutyrate, serotonin or dopamine is made in a short series of steps from an amino acid in the diet. For each step an enzyme in the cytoplasm of the cell acts as a catalyst. In general each nerve cell includes the enzymes that synthesize a single transmitter. The active form of the transmitter molecule is stored in the sacs called synaptic vesicles until the nerve cell is called on to release it. After the transmitter is released the cell can reabsorb some of it. In that way the need for the synthesis of the transmitter is somewhat reduced.

In a cell that releases a peptide the process is more involved. In the first place the peptide can be synthesized only by ribosomes, the specialized intracellular organelles that synthesize all peptides, including the long ones: proteins. This means the sequence of amino acids that make up the peptide must be encoded by a gene, a strand of DNA in the nucleus of the cell. The gene must be transcribed into a strand of messenger RNA, which carries the code to the ribosome. In the second place all the peptide neurotransmitters examined so far

appear to follow the pattern of the endorphins in that it seems they are synthesized first as a larger peptide. Then the active form of the molecule is produced through progressive cleavings by enzymes.

The ribosomes in a nerve cell lie only in the cell body and in the filamentous extensions of the cell body called dendrites. In general, however, the dendrites and the cell body of a nerve cell receive signals from other cells. A longer filament, the axon, transmits signals to other cells. Hence the release of a peptide neurotransmitter from an axon terminal is remote from the place where the peptide is made. The active form of the peptide (stored in synaptic vesicles) must be transported, then, to the place of release. It may follow that a cell releasing a peptide is unable to act on another cell repeatedly in a short span of time. In contrast, the axon terminals of a cell that releases an amino acid or a monoamine may well be satellite factories that make their own transmitter. Such terminals may stand less in need of replenished supplies of fresh neurotransmitter from moment to moment.

A further way in which a peptide neurotransmitter may differ from a monoamine neurotransmitter lies in the molecular details of how the transmitter influences its target cells. The classical neurotransmitters are said to be excitatory or inhibitory. The arrival of molecules of an excitatory neurotransmitter makes the target cell more likely to release its own neurotransmitter from its own axon terminals; the arrival of an inhibitory neurotransmitter has the opposite effect. Investigators who have studied these actions emphasize that the neurotransmitter alters the permeability of

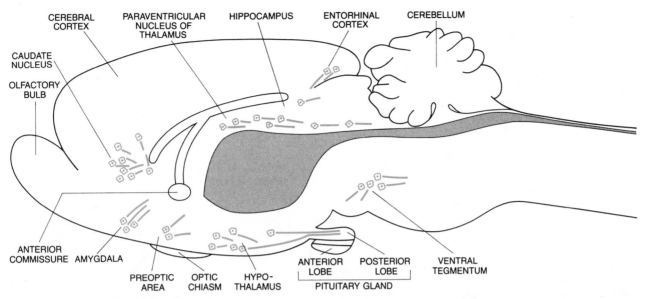

NERVE CELLS CONTAINING ENKEPHALIN are widespread in the brain of the rat, which makes them distinct from the nerve cells containing endorphin. Here the cells containing enkephalin are projected onto the midplane of the brain and displayed in proportion to the actual numbers of such cells at various locations. The trajectories of the axons arising from the cells are suggested by line segments.

the target cell's membrane to ions of potassium, sodium or calcium. As a result of such changes the concentration of these ions inside the target cell changes with respect to their concentration outside the cell. The differing concentrations give rise to a voltage gradient across the membrane. The action of an excitatory neurotransmitter tends to diminish the voltage gradient; it "depolarizes" the membrane. The action of an inhibitory transmitter tends to increase the voltage gradient; it "hyperpolarizes" the membrane.

In experiments with the enkephalins, substance *P* and other peptide messengers a different action appears. The peptide makes the target cell less likely to respond to other signals. One might consider this to be an example of inhibition. In some instances, however, the peptide messenger keeps an excitatory transmitter from depolarizing the membrane; in other instances it keeps an inhibitory transmitter from hyperpolarizing it. Moreover, the arrival of the peptide itself often produces no notable change in the voltage gradient across the membrane. This curious effect of a peptide messenger is one I like to call disenabling. It has now been reported by Roger Nicoll of the University of California at San Francisco, who studied peptides in the spinal cord of the frog; by Jeffrey L. Barker and his colleagues at the National Institute of Neurological and Communicative Disorders and Stroke, who studied enkephalin in the mammalian neurons they cultured from the spinal cord; by Walter Zieglgänsberger and his colleagues at the Max Planck Institute for Psychiatry, who studied enkephalin in the spinal cord of the cat, and by Zieglgänsberger and George R. Siggins at

the Salk Institute, who studied enkephalin in slices of hippocampus of the rat.

Since the peptide and the disenabled transmitter must each act on the target cell at a different set of receptors, one imagines the two sets of receptors might interact. Indeed, it seems that certain small populations of nerve cells can release both a monoamine and a peptide. Some are thought to release both acetylcholine and vasoactive intestinal peptide, others to release both dopamine and cholecystokinin and still others to release both serotonin and substance *P*. The arrival of such a pair of messengers at a target cell might give rise to a complex sequence of effects.

The Mediation of Behavior

The ultimate question about any intercellular messenger is how it integrates the activities of cells in a way that is appropriate to the circumstances in which the organism finds itself. Briefly, how does it mediate behavior? One answer is suggested by the work of Donald Pfaff and his colleagues at Rockefeller University and that of Robert L. Moss and Samuel M. McCann at the Southwestern Medical School of the University of Texas Health Science Center in Dallas. In each case the work concerns the decapeptide called luteinizing hormone releasing hormone (LHRH). LHRH is a substance purified from the hypothalamus that causes the release of luteinizing hormone from the anterior lobe of the pituitary. Luteinizing hormone in turn induces ovulation. LHRH has now been found in nerve cells in the autonomic ganglion that innervates the reproductive organs. Its utilization there instead of some other messenger may be

fortuitous. As it happens, however, the injection of LHRH into a male or female rat either subcutaneously or into a cerebral ventricle evokes the posture required for copulation. It is as if several activities that constitute reproductive behavior were coordinated by the same peptide messenger.

A further example of the mediation of behavior by a peptide is the ability of angiotensin II to bring on drinking behavior. The work of James T. Fitzsimons and his colleagues at the University of Cambridge shows that the injection of a few nanograms of angiotensin II under the skin or the injection of a few picograms into a cerebral ventricle causes behavior indistinguishable from spontaneous drinking in species ranging from lizards to primates. The animal's water balance or salt balance seems not to matter.

The discovery of angiotensin's behavioral effect adds to a considerable list of its physiological effects. Basically the events that activate angiotensin begin when any of three circumstances (low blood pressure, a low local concentration of sodium or direct stimulation by nerve fibers) causes certain cells in the kidney to secrete the enzyme renin. In the bloodstream renin acts on a protein manufactured in the liver and liberates a decapeptide called angiotensin I. In the bloodstream or in any of several organs including the brain a second enzyme cleaves two amino acids off from angiotensin I. An octapeptide remains. It is angiotensin II. Its physiological effects include the constriction of blood vessels supplying the skin and the kidneys, the dilation of blood vessels supplying muscles and the brain and an increase in blood pressure.

In addition angiotensin II causes the adrenal cortex to increase its output of aldosterone, a hormone that acts in turn on the kidneys to cause the reabsorption of sodium from the urine. Further still, angiotensin II increases the secretion of vasopressin from the posterior lobe of the pituitary. Vasopressin then promotes the reabsorption of water from the urine. All these actions tend to reverse the trends that triggered the secretion of renin. Thus they regulate three aspects of the blood: its volume, its pressure and its content of sodium. The fact that angiotensin II elicits drinking behavior is consistent with all its other effects.

According to the findings of Ian R. Phillips, who is now at the University of Florida, the cells in the brain that detect angiotensin are curious neurons in the hypothalamus and the brain stem that line the ventricles. Each one sends an extension of its cell body into a ventricle and yet each maintains typical synaptic connections with other neurons. The neurons they contact presumably include the neurons in the hypothalamus that secrete vasopressin. Remarkably, the subcutaneous or intraventricular injection of vasopressin does not elicit drinking behavior even when the experimental animal is dehydrated and presumably thirsty. On the other hand, the effects of circulating angiotensin do not account for every episode of drinking behavior. If they did, the behavior and the blood level of angiotensin would be linked. The work of Edward M. Stricker of the University of Pittsburgh suggests that this correlation is sometimes lacking. Doubtless the motivation to drink and the motor acts involved in seeking water and drinking it call for the simultaneous activity of several linked brain systems.

Although vasopressin does not cause animals to drink, it is reported to have effects on behavior that are even more dramatic. Over the past decade David de Wied and his colleagues at the State University of Utrecht and Abba J. Kastin and his colleagues at the New Or-leans Veterans Administration Hospital have reported observations that vasopressin and other peptides affect learning and memory. In one experimental arrangement rats were taught not to enter a dark box by being given an electric shock whenever they did. They were reported to continue to avoid the box for a longer period after the end of the training if they were given a subcutaneous injection of a minute amount of vasopressin. Molecular segments of corticotropin or of endorphin that have no known neural or endocrine actions were said to have the same property.

A Test of Vasopressin

Together with Michel Le Moal of the University of Bordeaux, George F. Koob and I set out three years ago to test the effects of vasopressin and other peptides on behavior. We resorted to an arrangement of de Wied's in which rats were trained to jump onto a pole whenever a light bulb lit up in warning that a painful electric shock was about to be delivered through the floor. When the training was complete, the light would be lit from time to time but no shock would be delivered. Before the training began we gave one group of rats an intraventricular injection of a saline solution; in general they stopped jumping onto the pole within four hours after the end of the training. We injected a second group of rats with a solution containing a nanogram of vasopressin; they continued to jump for eight to 10 hours after the end of the training. Strangely, a third group of rats, each given 50 nanograms of vasopressin, stopped jumping before the control group. It was as if the extra vasopressin had helped them to see through our ruse.

The change in behavior brought on by vasopressin seems, then, to be unquestionable. It is not yet clear, however, that the change represents the action of vasopressin on the unknown cellular processes that underlie learning and memory. The change may have a simpler explanation. In view of the effects of vasopressin on the circulatory system we were not surprised to find that even low doses of vasopressin caused brief but immediate elevations of blood pressure. We were sent a synthetic analogue of vasopressin by Sawyer of the Columbia College of Physicians and Surgeons and Maurice Manning of the Medical College of Ohio in Cincinnati. The analogue prevents vasopressin from elevating blood pressure. The rats to which the analogue was administered along with vasopressin showed no change in blood pressure, and they behaved like the control group.

The altered performance of the rats under the influence of vasopressin could mean, therefore, that the rats were aroused by an unnatural and unnecessary elevation of blood pressure and remained tense and alert for hours. The rats were aroused, that is, by a change in the body that the brain did not request. This seems less impressive than the hypothesis that vasopressin mediates learning and memory directly. On the other hand, a mismatch between the brain's commands and the body's responses may be sufficient to invoke behavioral strategies for dealing with novel situations. People given long-lasting analogues of vasopressin report enhanced attention to their surroundings, and their performance on tests of memory improves.

Moreover, even on the simpler hypothesis a neuropeptide apparently serves to signal that the survival of the animal is challenged and that the animal had best be attentive to its surroundings. Conversely, the absence of such a signal may suggest that the animal is momentarily safe. Survival signals such as these could represent a means by which the neuropeptides guide the evolution of complex forms of behavior. Surely the various aspects of research on peptide messengers are likely to advance understanding both of cellular regulation and of the means by which certain types of animal behavior result from decipherable cell-to-cell interactions.

5

Nutrients That Modify Brain Function

By Richard J. Wurtman
April 1982

*They are the precursors of neurotransmitter molecules.
Increasing their level in the brain amplifies signals
from some nerve cells. In effect they act like drugs,
and one day they may serve as drugs*

A nutrient is different from a drug, most people would agree. A nutrient is a food substance that in most cases supplies either the energy or the molecular building blocks the body requires. A drug is a substance given for its effect on a specific organ or type of cell. Whereas all healthy people need essentially the same nutrients, a drug would ordinarily be recommended only for people with a particular disease or condition. In this article I shall tell about three nutrients that, when they are administered in the pure form or simply ingested in food, can act like drugs. They give rise to important changes in the chemical composition of structures in the brain. The changes can modify brain function, particularly in people with certain metabolic or neurologic diseases.

Two of the nutrients are the amino acids tryptophan and tyrosine. Amino acids are the building blocks of proteins, and so tryptophan and tyrosine are present in most foods. The third nutrient is choline, a component of lecithin; egg yolks, liver and soybeans are notably rich in lecithin. The composition and function of the brain can be altered by tryptophan, tyrosine and choline because they are the precursors of neurotransmitters: substances that are released from a neuron, or nerve cell, when it fires. The neurotransmitter thereby conveys the nerve impulse across a synapse to either another neuron, a muscle cell or a secretory cell. Tryptophan is converted in the terminals of certain neurons into the neurotransmitter serotonin. In other cells choline is converted into the transmitter acetylcholine. In still another population of cells tyrosine serves as the precursor of dopamine, norepinephrine and epinephrine, which are collectively called the catecholamine transmitters. An increase in the brain level of a precursor enhances the synthesis of the corresponding neurotransmitter product. The enhanced synthesis can in turn cause the neuron to release more transmitter molecules when it fires, amplifying the transmission of signals from the neuron to the cells it innervates.

My associates and I at the Massachusetts Institute of Technology and other investigators have been exploring the interactions that relate the amount of a nutrient administered or ingested to its level in the blood plasma, its level in the brain and its effect on nerve transmission. The interactions are not simple. The conversion of tryptophan into serotonin is influenced by the proportion of carbohydrate in the diet; the synthesis of serotonin in turn affects the proportion of carbohydrate an individual subsequently chooses to eat. In the case of choline and tyrosine the effect on a neuron of an increased supply of the nutrient both varies with the neuron's firing frequency and can lead to changes in that frequency. Choline and tyrosine can therefore amplify neurotransmission selectively, increasing it at some synapses but not at others. It may be possible to exploit such selectivity to develop novel therapeutic agents for several diseases, including hypertension, some forms of depression, Parkinsonism and some memory disorders of old people.

The observations relating nutrient intake to neurotransmission originated in studies of a phenomenon seemingly unrelated to the brain: daily rhythms in the metabolism of dietary amino acids. When people eat, the concentration in their blood plasma of most amino acids (and of other food constituents) changes predictably in ways that depend on what foods are eaten. For people who take their meals at the usual times plasma amino acid levels generally exhibit pronounced daily rhythms. For example, among people consuming the high-protein diet typical of the U.S. the plasma concentration of the amino acid leucine is twice as high between 3:00 P.M. and 3:00 A.M. as it is during the rest of the day. If the same people eat protein-free meals, the leucine level instead falls by half during these hours of active digestion and absorption. In the first case the increase represents the entry into the bloodstream of some of the leucine in the dietary protein. In the second case the decrease results from the secretion of insulin (induced by ingested carbohydrate), which accelerates the passage of leucine and most other amino acids from the circulation into skeletal muscle.

My associates and I discovered these rhythmic, food-induced variations in the plasma level of various nutrients about a decade ago. We wondered whether the changes might have any functional significance. In particular we wondered whether the fluctuations in the concentration of circulating nutrients had any effect on the rate at which the nutrients are converted into cellular constituents. In order for changes in the level of a nutrient to influence the rate of conversion, the enzyme catalyzing the conversion must have a certain property. The enzyme's ability to bind the nutrient preparatory to changing its chemical structure must be relatively poor, so that at the usual nutrient concentrations each enzyme molecule is less than fully saturated with the nutrient and functions at less than peak efficiency. In this situation the quantity of the nutrient available to the enzyme is the rate-limiting element in the reaction, and so an increase in the nutrient's concentration increases the level of enzyme activity: more of the nutrient is converted and more of the product is formed.

We knew that tryptophan is converted into serotonin by just such a low-affinity enzyme. The release of serotonin by neurons originating in the brain stem delivers signals to widely scattered groups of neurons that control such things as sleep, mood and appetite. Other investigators had already shown that the concentration of serotonin in the brain can be increased by giving experimental animals very large doses of pure tryptophan. John D. Fernstrom and I decided to examine the possibility that normal daily variations in the plasma concentration of tryptophan might be enough to alter the rate of serotonin synthesis in the rat brain. We found that even low doses of tryptophan, which raise the plasma concentration of the amino acid but keep it within the normal daily range we had previously established, did indeed enhance serotonin synthesis.

To see whether a reduction in plasma tryptophan had the opposite effect we injected some rats with insulin and gave others a diet of carbohydrates, which induces the secretion of insulin. We expected that the hormone would reduce the plasma level of tryptophan as it does that of other amino acids by moving them out of the bloodstream and into skeletal muscle. To our surprise the insulin did not lower the tryptophan concentration in the plasma, and it actually raised the concentration in the brain, increasing serotonin synthesis instead of reducing it. Feeding the animals large amounts of protein brought another surprise: even though amino acids were plentiful in the diet, both the brain concentration of tryptophan and the synthesis of serotonin were reduced.

The apparent paradoxes were resolved when we found that the amount of tryptophan available in the brain for conversion into serotonin depends not only on the amount of tryptophan in the plasma but also on the ratio of plasma tryptophan to the plasma level of five other amino acids: tyrosine, phenylalanine, leucine, isoleucine and valine. All six of these amino acids are comparatively large molecules, and in a physiological environment most of them are electrically neutral, with about as much positive charge as negative.

It is very difficult for large, water-sol-

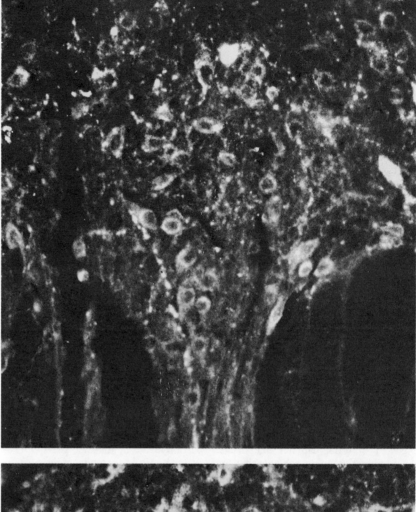

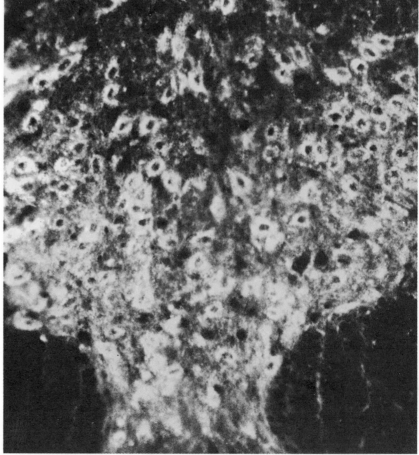

SYNTHESIS OF SEROTONIN, a neurotransmitter, is greatly increased by the administration of the nutrient that is its precursor: the amino acid tryptophan. These photomicrographs, made by George K. Aghajanian of the Yale University School of Medicine, show neurons (nerve cells) in the dorsal raphe nucleus of the rat, one of the structures where most of the brain's serotonin is synthesized and stored. Thin brain-tissue sections were treated with formaldehyde, which reacts with serotonin to form a greenish yellow fluorescent compound. In the control section (upper photograph) faint fluorescence reveals the presence of serotonin in the cytoplasm of the cell bodies and along the axons, or nerve fibers. The other section (lower photograph) is from the brain of an animal that was injected with a large dose of tryptophan an hour before the tissue sample was taken. There is a dramatic increase in the level of fluorescence, reflecting an increased synthesis of serotonin in the presence of the additional precursor.

uble molecules to diffuse out of the capillaries of the brain and gain access to neurons and other brain cells. Their passage between the blood and the brain is facilitated by carrier molecules present in the endothelial cells lining brain capillaries. A single species of carrier molecule transports all six of the large, neutral amino acids across the blood-brain barrier; the amino acids compete with one another for attachment to the carrier and hence for uptake from the bloodstream into the brain. There is far less tryptophan in most proteins than there is tyrosine, phenylalanine, leucine, isoleucine or valine. A high-protein meal therefore reduces the plasma ratio of tryptophan to the competing amino acids; less tryptophan is carried across the barrier and less reaches the neurons.

A high-carbohydrate meal has the opposite effect because the insulin secreted in response to carbohydrate intake reduces the plasma level of the competing amino acids more than it does that of tryptophan. Whereas the other amino acids circulate as free molecules, most of the tryptophan is bound to the plasma protein albumin; segregated in an albumin reservoir, the tryptophan is essen-

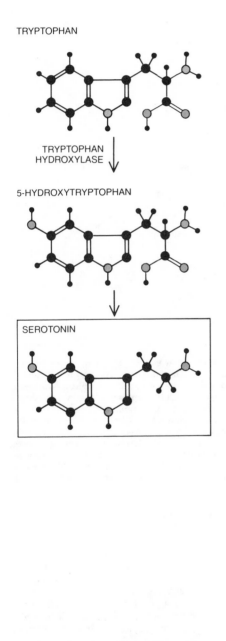
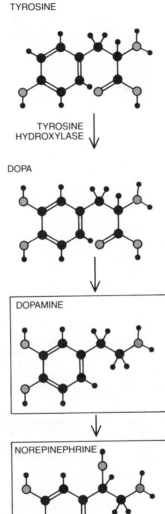
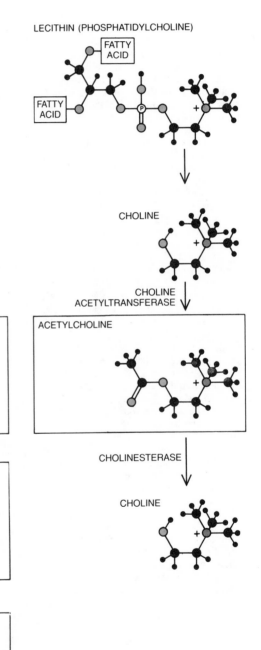

- ● CARBON
- • HYDROGEN
- ◐ OXYGEN
- ◑ NITROGEN
- Ⓟ PHOSPHORUS

CHEMICAL STRUCTURE of three nutrients and the paths by which they are converted into neurotransmitters are shown. Tryptophan is converted into serotonin in two steps. The three catecholamine transmitters, dopamine, norepinephrine and epinephrine, are formed in successive steps from another amino acid: tyrosine. The food substance lecithin provides choline, the precursor of the transmitter acetylcholine. Unlike other precursors, choline is recycled; it is regenerated when acetylcholine is broken down by cholinesterase.

tially immune to the effect of insulin. The result is that after carbohydrates are eaten the plasma ratio of tryptophan to its competitors rises, causing more tryptophan to reach the neurons [see "Nutrition and the Brain," by John D. Fernstrom and Richard J. Wurtman; SCIENTIFIC AMERICAN Offprint 1291].

We proposed that these interactions enable the serotonin-releasing neurons in the brain to serve as sensors of the plasma tryptophan ratio, which increase serotonin release after a carbohydrate meal and reduce it after a high-protein meal. David Ashley and G. Harvey Anderson of the University of Toronto Faculty of Medicine subsequently found evidence that the brain exploits this property of the serotoninergic neurons when an animal chooses one food in preference to another. In an effort to define the specific food constituents whose choice is influenced by brain serotonin, Judith J. Wurtman and I allowed rats to choose between two diets having different proportions of carbohydrate and protein. Various treatments that increase serotonin release in the brain (such as giving the drug fenfluramine) caused the rats to selectively reduce their consumption of carbohydrate. Recently we have shown that serotonin-increasing treatments have a similar effect on obese people with a craving for dietary carbohydrate when they are allowed to choose from a range of snack foods over a period of several weeks.

It appears, in other words, that eating a meal rich in carbohydrate and poor in protein generates a neurochemical change—namely increased serotonin synthesis—that causes the animal to reduce its intake of carbohydrate but not of protein. It seems likely that this control of serotonin release by diet composition and of diet composition by serotonin release evolved because it helps to sustain nutritional balance. Presumably it keeps the bear from eating only honey and keeps human beings from eating sweets and starches to the exclusion of enough protein. Some obese people may suffer from a disturbance of this remarkable feedback mechanism that interlinks nutritional, metabolic, neurochemical and behavioral systems. Unfortunately there is no noninvasive technique for measuring serotonin release in the brain and thus directly establishing the role of serotonin release in man.

In 1975 Edith L. Cohen and I, and independently Dean R. Haubrich of the Merck Institute for Therapeutic Research, showed that the administration of choline increases the synthesis of acetylcholine in the brain. (Consumption of lecithin, the food constituent that provides most of the choline in adult diets, is even more effective.) Soon after that Candace J. Gibson and I observed a similar relation between meal composition and the synthesis of the catecholamine

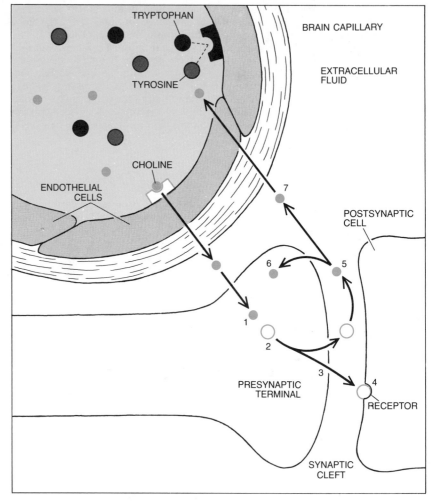

NUTRIENT MOLECULES must cross the "blood-brain barrier" to get to the brain cells where they serve as precursors of neurotransmitters. The junctions between the endothelial cells of the capillaries in the brain are too tight to allow the passage of tryptophan, tyrosine or choline, and so the precursors must be transported through the capillary wall by carrier molecules. Tryptophan, tyrosine and other large, electrically neutral amino acids compete for the same carrier. The uptake and conversion of choline are shown here. Choline in the extracellular fluid of the brain is taken up by the terminal of a cholinergic neuron (1) and is converted into acetylcholine (2). The acetylcholine is released into the synaptic cleft when the neuron fires (3). The acetylcholine may interact with a receptor and thereby transmit a signal to the postsynaptic cell (4). Alternatively, the transmitter may be converted back into choline (5), which may be taken up again by presynaptic terminal (6) or may enter extracellular fluid and bloodstream (7).

transmitters dopamine, norepinephrine and epinephrine. The administration of tyrosine or the consumption of meals that increase the relative plasma level of tyrosine (the level in relation to the other large, neutral amino acids) raises the tyrosine level in neurons, thereby making more of the amino acid available to the enzyme tyrosine hydroxylase and accelerating catecholamine synthesis.

With three nutrients shown to affect the synthesis of their neurotransmitter products, it became possible to formulate some general principles for predicting whether a given transmitter might be subject to such control and to propose some standardized experiments for demonstrating the existence of such relations. I can illustrate the principles by describing the nature of the five biochemical processes that must follow in sequence if the consumption of a meal

rich in a nutrient is to increase the synthesis in the brain of the transmitter for which the nutrient is the precursor.

First, consumption of a food that includes the nutrient must significantly elevate the plasma level of the nutrient; the plasma level cannot be held relatively constant by feedback mechanisms like the ones that regulate plasma pH, say, or the concentration of calcium. Second, the concentration of the nutrient in the brain must depend on and vary with the concentration in the plasma; there cannot be an impenetrable blood-brain barrier for the precursor. Third, the transport mechanism that mediates the nutrient's movement between the blood and the brain must be of the low-affinity type: it must be unsaturated with its substrate (the precursor), so that it can become more nearly saturated when the plasma level rises. Fourth, the neuronal enzyme that catalyzes the conversion of the precursor into the transmitter must also be a low-affinity one. (As I mentioned above, it was this requirement that led us to the effect of tryptophan on serotonin-releasing neurons.) Fifth, the enzyme must not be susceptible to feedback inhibition when the intracellular level of its product, the transmitter, rises.

All these conditions have now been shown to be fulfilled in the synthesis of serotonin, acetylcholine and the catecholamines. There is some evidence that they are also met in the case of two other transmitters, histamine and glycine, whose production seems to be affected by the availability of their precursors. The fact that so many conditions must be met if precursor availability is to influence neurotransmitter synthesis implies that the precursor-transmitter relation is no biological accident. Rather it seems likely the relation has some adaptive value and has therefore been conserved in evolution.

Are all neurotransmitters subject to precursor control? Probably not. The immediate precursors of some transmitters are usually available in concentrations that fully saturate the neurotransmitter-synthesizing mechanisms and are therefore independent of plasma composition. In other cases it is not possible to assess the extent to which precursor levels control the synthesis of a transmitter, either because the precursor's identity remains to be established or because blood-brain transport systems frustrate the experimenter's attempts to raise the concentration of the precursor in the brain. Even if only some of the 25 or so neurotransmitters known are potentially responsive to nutrient consumption, however, it is interesting that the responsive transmitters include several whose action is thought to be affected by drugs given to treat neuro-

logic, psychiatric and even cardiovascular diseases.

Although the intake of a nutrient may influence the synthesis of a neurotransmitter in the presynaptic terminal of a neuron, it does not necessarily follow that the nutrient alters the transmission of impulses across the synapse. The nutrient must also be shown to increase the number of transmitter molecules released by a given set of neurons per unit time. The number depends mainly on the number of neurons in the tract or nerve being examined, on the total number of synapses they make, on the frequency with which the neurons fire and on the average number of molecules released at each synapse each time the neurons fire. It was theoretically possible that nutrient administration affected only the synthesis of a transmitter and not its release into synapses. This was shown not to be the case in experiments done by George G. Bierkamper and Alan M. Goldberg of the Johns Hopkins University School of Medicine. They measured the release of acetylcholine from motor neurons incubated in various choline concentrations. When they stimulated the nerves electrically, they found a striking parallel between the concentration of choline and the amount of acetylcholine released.

Tracing the effect of added nutrient one step further, Ismail Ulus of the University of Bursa in Turkey and I showed that the increase enhances not only transmitter release but also neurotransmission, giving rise to a chemical change in the postsynaptic cells. In these studies we worked with the splanchnic nerve, which runs from the spinal cord to the adrenal glands. There the splanchnic-nerve fibers terminate in synapses with cells called chromaffin cells in the medulla, or core, of the glands. When the splanchnic nerve fires, it releases acetylcholine. This transmitter causes the postsynaptic chromaffin cells to release epinephrine and also to make more tyrosine hydroxylase, the enzyme that ultimately controls the synthesis of epinephrine.

Other investigators (notably Julius Axelrod, Hans Thoenen and Robert A. Mueller, who were all then at the National Institute of Mental Health) had shown that if rats were given a treatment that chronically increased the firing rate of the splanchnic nerve, the consequent release of added acetylcholine increased the level of tyrosine hydroxylase activity in the adrenal gland. We thought that giving choline would increase the amount of acetylcholine released per firing without altering the splanchnic nerve's firing frequency and therefore might similarly increase tyrosine hydroxylase activity. It did. Then we tested animals in which the splanchnic-nerve fibers supplying one of the two adrenal glands had been severed.

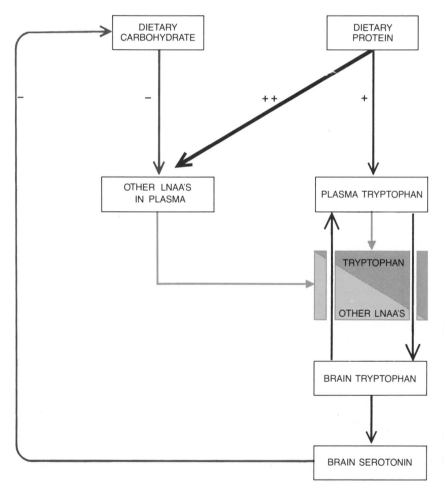

FOOD CONSUMPTION affects the synthesis of serotonin in the brain. Eating protein raises the plasma level of tryptophan, but it raises the plasma level of five other large, neutral amino acids (LNAA's) even further because each of them is more plentiful than tryptophan in proteins. Dietary carbohydrate induces the secretion of insulin, which moves most amino acids out of the bloodstream but has little effect on tryptophan. Because tryptophan must compete with the other LNAA's for transport across the blood-brain barrier, the movement of tryptophan from the plasma to the brain is controlled by the plasma ratio of tryptophan to the other LNAA's (color). When the ratio is high, tryptophan enters the brain; when it is low, tryptophan moves from the brain into the bloodstream. The release in the brain of serotonin synthesized from the tryptophan appears to reduce a rat's (or a person's) carbohydrate intake.

We found that choline induced greater enzyme activity in the intact gland but failed to do so in the denervated one, proving that circulating choline did not act directly. It needed first to be converted into acetylcholine, which was released from the splanchnic nerve.

To confirm that the choline acted by changing the amount of neurotransmitter released per firing (and not by raising the firing frequency) we tested the combined effect of giving rats both choline and a treatment meant to accelerate splanchnic-nerve firing. (The treatments included putting animals in a cold environment, giving them very large doses of insulin and administering drugs that cause prolonged depression of blood pressure.) We reasoned that if choline and the other treatment both acted by increasing the firing frequency, their combined effect on tyrosine hydroxylase would be about the same as the effect of one treatment alone. If, on the other hand, choline increased the amount of acetylcholine released per firing, the combined effect should be multiplicative. We observed that the effects of giving choline and of the second treatment were indeed multiplicative. To our surprise, however, the two treatments also potentiated each other, that is, their combined effect on tyrosine hydroxylase activity was invariably larger than the product of their effects when they were given alone.

These experiments led to two conclusions. One was that giving a nutrient molecule that acts as a precursor can enhance the release of a neurotransmitter and thereby amplify a neuron's effect on postsynaptic cells. In the instance studied the administration of choline leads to the release of more acetylcholine and hence to greater tyrosine hydroxylase activity. The other conclusion, which had not been expected, was that the magnitude of choline's effect varies with the firing frequency of the neurons on which it acts. The latter finding provided the first suggestion of the important relation, which I mentioned at the outset, between the firing frequency of a neuron and the extent to which it responds to an increased supply of its transmitter's precursor.

Within four months of the publication of our first article describing the increase in brain acetylcholine in rats given choline, the first clinical application of the nutrient-neurotransmitter relation was reported. The report concerned a patient suffering from tardive dyskinesia, a disease characterized by uncontrollable movements of the face and upper body; it is caused (in a substantial fraction of patients) by the prolonged administration of antipsychotic drugs and is considered by many psychiatrists to be the most important side effect limiting the use of those drugs.

Kenneth L. Davis and his colleagues

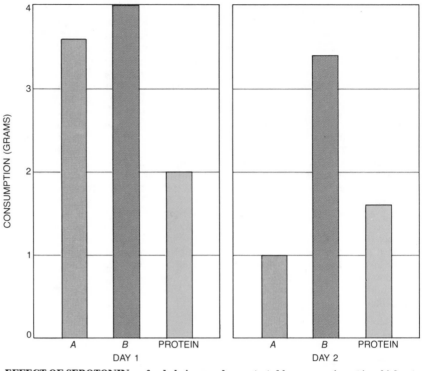

EFFECT OF SEROTONIN on food choice was demonstrated by an experiment in which rats were offered their choice of two diets. Each diet provided the same amount of carbohydrate and the same number of calories, but diet *A* was low in protein (5 percent), whereas diet *B* was protein-rich (45 percent). On the first day the rats were injected with a placebo. On the second day they were injected with fenfluramine, a drug that increases serotonin release. The additional serotonin caused a selective reduction in the amount of food consumed from diet *A*, so that the rats' carbohydrate intake decreased but protein intake stayed about the same.

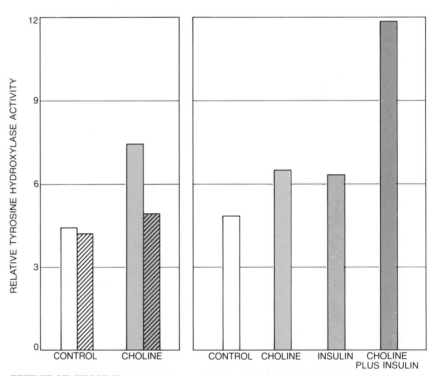

EFFECT OF CHOLINE on signal transmission by acetylcholine from splanchnic-nerve terminals to the chromaffin cells of the adrenal gland was assessed by measuring the activity of tyrosine hydroxylase, an enzyme whose synthesis is augmented by acetylcholine. In one experiment (*left*) one of the two adrenal glands was denervated (*hatched bars*). Giving choline for several days significantly increased enzyme activity in the intact gland but not in the denervated one. In a second experiment (*right*) two treatments were tested in rats, first alone and then in combination. Giving choline increased enzyme activity; so did the injection of insulin, which accelerates the firing of the splanchnic nerve. The two treatments potentiated each other: their combined effect was more than twice as great as the sum of their individual effects.

at the Stanford University School of Medicine had found that giving patients physostigmine, a drug that increases acetylcholine levels (by blocking the degradation of the transmitter by the enzyme cholinesterase), could temporarily ameliorate the abnormal movements. They then showed that increasing brain acetylcholine by giving choline has a similar effect. Other investigators have since confirmed the efficacy of choline in many patients with tardive dyskinesia, including a large group of people treated in a double-blind, placebo-controlled test by my collaborator John H. Growdon of the Tufts University School of Medicine.

Choline has largely been replaced by lecithin as the precursor of choice for raising acetylcholine levels in patients because lecithin is less susceptible to bacterial degradation in the intestine. The degradation wastes the choline and also leads to the formation of trimeth-

ylamine, which can impart the aroma of rotten fish to the hapless patient. Unfortunately "lecithin" means a specific chemical compound, phosphatidylcholine, to physicians and biochemists, whereas in the food industry it encompasses an entire family of substances, the phosphatides. Only phosphatidylcholine is an effective source of choline; virtually all the lecithin preparations sold in health-food stores are too impure to be of much value to patients.

The administration of choline or lecithin does not interfere with the therapeutic effects of antipsychotic drugs, only with their side effect, tardive dyskinesia. Moreover, unlike physostigmine and other inhibitors of cholinesterase, choline and lecithin do not give rise to the side effects associated with increased acetylcholine activity itself. The amplified acetylcholine transmission that follows the administration of physostigmine, for example, leads to excessive

formation of mucus in the respiratory passages, a very low heart rate, gastrointestinal cramping and numerous alterations of brain function. What explains the absence of such cholinergic side effects after the administration of lecithin or choline?

The answer goes back to the interdependence of firing frequency and neurotransmitter synthesis [see illustration below]. Not all neurons that are potentially able to release more transmitter molecules in the presence of more precursor actually do so. The brain can cause particular groups of normally functioning neurons to diminish their firing frequency, so that although they synthesize more transmitter molecules, they do not release more molecules per unit time. The diminished firing frequency has a second effect: it somehow reduces the sensitivity of the slower-firing neurons to the presence of additional precursor. In tardive dyskinesia the neurons that

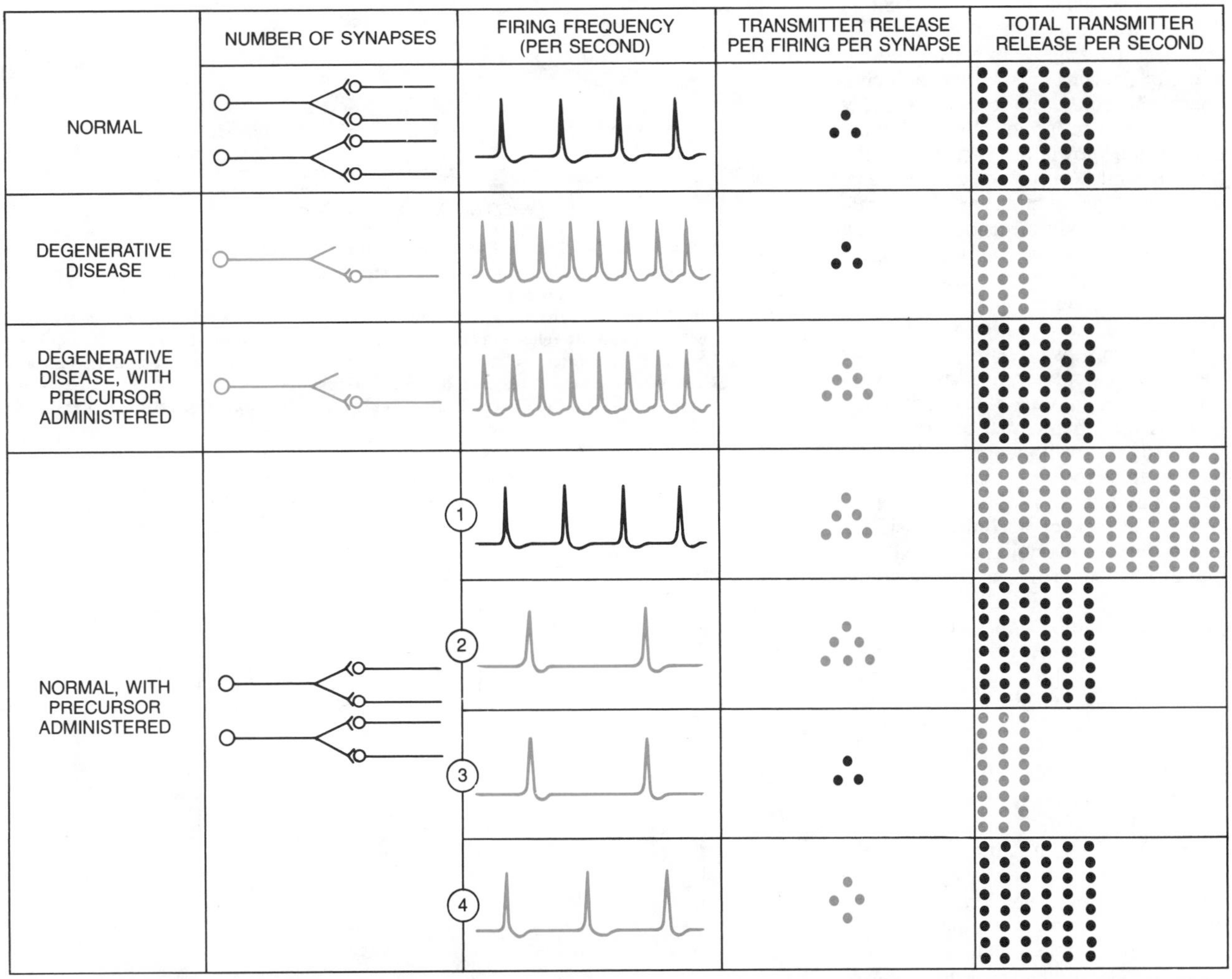

FIRING FREQUENCY AND PRECURSOR SUPPLY are interrelated. A hypothetical set of four normal synapses is shown at the top; the neurons fire four times per second, and at each synapse three transmitter molecules are released per firing. The result is a total release of 48 molecules per second. A hypothetical degenerative disease reduces the number of neurons and synapses. The remaining neuron fires faster, but transmitter release is still inadequate. The administration of additional precursor increases the number of transmitter molecules released at each firing and thus achieves a normal total rate of release. If additional precursor is supplied to a normal set of neurons, the response is different. First there is a transient increase in total release because more transmitter is synthesized and more is released at each firing (1). Soon, however, the neurons' firing rate is slowed, reducing transmitter release to the normal level (2). The reduced rate in turn somehow reduces the neurons' sensitivity to the additional precursor; less transmitter is released per firing (3). Eventually a combination of firing rate and release per firing is reached that maintains normal total rate of transmitter release (4).

suppress the symptoms are apparently sensitive to the additional supply of
choline, and so their action is amplified. Other cholinergic neurons first slow
their firing and then become unresponsive to additional choline; as a result
their release of acetylcholine does not
increase and the side effects associated
with overstimulation of the cholinergic
neurons are absent.

Any disease state known to result from
inadequate cholinergic neurotransmission, in any part of the body, now
becomes a candidate for treatment with
lecithin given either alone or as an ad-
junct to drug therapy. A number of brief
reports have described improvement after lecithin administration to individual
patients suffering from such diseases.
Except in the case of tardive dyskinesia,
however, too little information is available to sustain even tentative conclusions as to lecithin's therapeutic efficacy.

The diseases currently generating the
most interest as candidates for lecithin
therapy are the memory disorders associated with old age. Aging brings with it
a loss of neurons in the brain, and cholinergic neurons seem to be particularly
vulnerable. The hippocampus, a region
of the brain known to be essential for the
formation of new memories, has a particularly large number of cholinergic
neurons. The administration to young
people of drugs such as scopolamine,
which block cholinergic transmission,
causes short-term memory impairments
similar to those observed in the aged.
For these reasons it seems possible that
treatments calculated to increase brain
acetylcholine may be effective in some
patients with memory disorders.

Investigators who would study possible therapies for these disorders are be-
set by a number of problems. One problem is the lack of a clinical basis for
classifying memory-impaired patients
into distinct subgroups according to the
origin of the disability. It seems likely
that the population of patients now said
to have senile dementia or Alzheimer's
disease will one day be found to consist
of people with several different diseases,
only some of which may reflect a selective decrease in brain acetylcholine and
so perhaps be amenable to choline therapy. For now there is no way to distinguish such patients.

Another problem is the lack of well-
validated tests for measuring improvement in memory functions. There are
objective tests of memory, but in no case
has it been possible to demonstrate that
an improvement in a test score presages
improvement in the patient's real-life
memory functions. The tests cannot be
validated because (it is a vicious circle)
no treatment has yet been discovered
that demonstrably improves real-life
memory functions. Furthermore, the
substances available for testing are less
than satisfactory. Most partially puri-

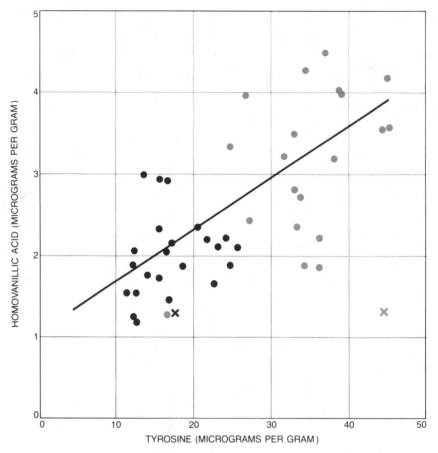

ADMINISTRATION OF TYROSINE leads to the accumulation of homovanillic acid, a
breakdown product of dopamine, in the corpus striatum of the rat, but only when the firing
of dopamine-releasing neurons is speeded up. Giving tyrosine alone (*colored* X) increases the
concentration of tyrosine in the brain compared with the level in control animals (*black* X), but
it has no effect on dopamine release (as measured by the homovanillic acid level). Giving haloperidol, a drug that accelerates the firing of the neurons, enhances the release of dopamine and
also makes the synthesis and release of dopamine dependent on the brain level of tyrosine
(*black dots*); now administration of tyrosine markedly increases dopamine release (*colored dots*).

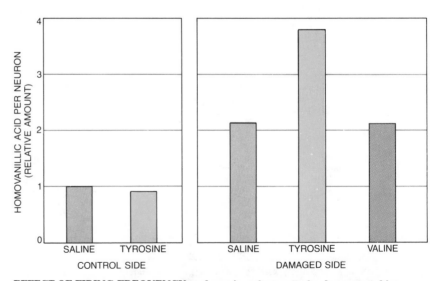

EFFECT OF FIRING FREQUENCY on dopamine release was also demonstrated by an experiment in which 75 percent or more of the dopamine-releasing neurons on one side of the rat
brain were destroyed, causing the remaining neurons to fire more rapidly. The injection of tyrosine rather than a presumably inert saline solution does not significantly change the amount of
dopamine released per neuron on the intact side of the brain. On the damaged side the increased firing rate of the remaining neurons is manifest: in animals given saline solution the
amount of dopamine released per neuron is about doubled. Giving tyrosine leads to a further increase in dopamine release; giving valine, another large, neutral amino acid, does not.

fied lecithin preparations either contain too little phosphatidylcholine or are unpalatable in the large doses that seem to be necessary. The standard drugs with which investigators enhance cholinergic function in animals act too nonspecifically on cholinergic synapses everywhere (and so cause too many side effects) for testing in human beings.

The few reports that have appeared on the memory responses of patients given lecithin suggest that it can be helpful in some cases but is not by any means a fountain of youth. Perhaps combinations of lecithin with drugs that accelerate the firing of cholinergic neurons in the hippocampus or with low doses of cholinesterase inhibitors will be useful in treating patients who have memory disorders attributed to a localized cholinergic deficit. Growdon and Suzanne H. Corkin of M.I.T. are currently testing several such combinations.

The close relation between the firing frequency of a neuron and its response to an increased supply of its transmitter's precursor is best illustrated by a group of dopamine-releasing neurons, those that run from the substantia nigra in the midbrain to the corpus striatum deep within each cerebral hemisphere. These nigrostriatal neurons participate in the control (particularly in the initiation) of movements and are severely damaged in Parkinson's disease.

The release of dopamine from the nigrostriatal neurons can be assessed in animals by measuring the accumulation

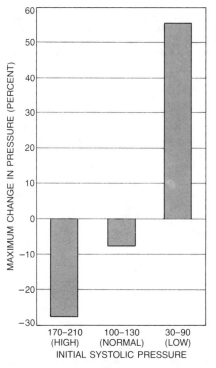

EFFECT OF TYROSINE on blood pressure depends on the initial pressure. The amino acid lowers the blood pressure of hypertensive rats and raises it in hypotensive animals.

in the corpus striatum of a dopamine metabolite, homovanillic acid. If otherwise untreated animals receive even a large dose of tyrosine, the nutrient that is the precursor of dopamine, no change is noted in the concentration of homovanillic acid in the corpus striatum. If the animals are pretreated with a drug that accelerates the firing of the nigrostriatal neurons, however, dopamine release becomes sensitive to the tyrosine level.

This effect was demonstrated elegantly by Franz Hefti and Eldad Melamed (who at the time were working with me at M.I.T. but who are now respectively at the Sandoz Research Laboratories in Basel and the Hebrew University of Jerusalem Faculty of Medicine). By administering a modified form of dopamine that damages dopaminergic neurons, they disabled more than 75 percent of the nigrostriatal neurons in one hemisphere of the rat brain. The surviving neurons in the damaged hemisphere responded by increasing their firing frequency; they are probably stimulated to do so by a feedback mechanism that enables nigrostriatal neurons to take over one another's functions. (A similar compensatory process keeps most Parkinsonian patients from showing symptoms until at least half of their nigrostriatal neurons have been lost.) The accelerated firing was confirmed by showing that the homovanillic acid level per surviving neuron was much higher on the damaged side of the brain than on the intact side. The animals then received tyrosine. On the damaged side there was a further increase in the homovanillic acid level per neuron; on the intact side there was no increase.

Tyrosine in the diet can also either increase the synthesis of norepinephrine or leave it unchanged depending on the firing frequency of particular neurons. This property explains the remarkable effects of tyrosine on blood pressure. When tyrosine is administered to animals (or people) with normal blood pressure, it has no consistent effect on the pressure. When tyrosine is given to hypertensive rats, however, it markedly lowers their blood pressure; when it is given to animals with hypotension (animals in shock, for example), it raises their blood pressure to near-normal levels. These observations were made by my graduate students Alan F. Sved, Lydia Conlay and Timothy J. Maher, who worked with rats that were genetically hypertensive and with other rats whose blood pressure had been lowered by reducing their blood volume by about 20 percent.

In a hypertensive animal the brain, acting to reduce blood pressure, accelerates the firing of norepinephrine-releasing neurons in the brain stem. At this site norepinephrine acts as an inhibitory neurotransmitter: it suppresses the firing of other neurons, ultimately diminishing the activity of the peripheral sympathetic neurons and the chromaffin cells of the adrenal medulla. The sympathetic neurons and chromaffin cells therefore liberate less norepinephrine and epinephrine, which would otherwise act to raise blood pressure by causing blood vessels to constrict and cardiac output to rise. Since only the brain-stem neurons are firing frequently, only their inhibitory output is amplified by tyrosine administration, and so the blood pressure falls. In animals in shock, on the other hand, the brain acts to raise the blood pressure: the inhibitory norepinephrine-releasing neurons of the brain stem are suppressed, whereas the sympathetic neurons and chromaffin cells are activated to fire more frequently. In this instance tyrosine selectively enhances catecholamine release from those cells, and the blood pressure rises.

Because of this regulatory mechanism tyrosine has a distinct theoretical advantage over many of the drugs now favored for the treatment of circulatory disorders. In theory, at least, it can be expected to act without overshooting the mark because as soon as normal blood pressure is attained, selective changes in neuronal firing frequencies should render the animal insensitive to the effects of additional tyrosine. Tyrosine has yet to be systematically tested in people with hypertension or shock, however, and theoretical arguments are no substitute for well-designed tests of clinical efficacy and safety.

Another human disease state in which the possible therapeutic value of tyrosine is currently being tested at several institutions is depression. Most psychiatrists who seek biochemical explanations of mental illness think that in many patients depression reflects inadequate neurotransmission mediated by either norepinephrine or serotonin. If norepinephrine release is inadequate in certain regions of the brain of some depressed patients, the administration of tyrosine could conceivably be helpful to them. The first evidence that tyrosine can have an antidepressant effect was obtained in studies we did in collaboration with Alan J. Gelenberg of the Massachusetts General Hospital; the findings were quickly confirmed elsewhere. If further clinical testing shows tyrosine can help in treating depression, it will be interesting to learn how it does so. In patients who respond to tyrosine is something wrong with the metabolism of the amino acid (causing the plasma tyrosine ratio to be too low), or is the fault perhaps in the conversion of tyrosine into norepinephrine in brain neurons?

When large doses of a nutrient, separated from the other constituents of foods that are its usual source, are given to people specifically to treat a disease or condition, does the nutrient thereby become a drug? The question is not merely one of nomenclature. The

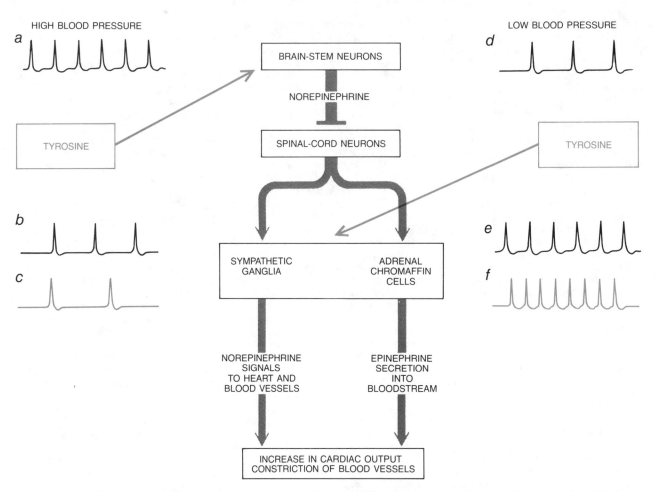

HIGH BLOOD PRESSURE

LOW BLOOD PRESSURE

a

d

BRAIN-STEM NEURONS

NOREPINEPHRINE

SPINAL-CORD NEURONS

TYROSINE

TYROSINE

b

c

e

f

SYMPATHETIC GANGLIA

ADRENAL CHROMAFFIN CELLS

NOREPINEPHRINE SIGNALS TO HEART AND BLOOD VESSELS

EPINEPHRINE SECRETION INTO BLOODSTREAM

INCREASE IN CARDIAC OUTPUT CONSTRICTION OF BLOOD VESSELS

ABILITY OF TYROSINE to both reduce high blood pressure and raise low blood pressure is explained by the effect of the amino acid on different populations of cells. In response to high blood pressure norepinephrine-releasing neurons in the brain stem speed their firing (*a*). The effect of the norepinephrine released by these neurons is inhibitory: by suppressing the firing of preganglionic neurons of the sympathetic nervous system in the spinal cord it reduces the activity of neurons in the sympathetic ganglia and of the adrenal chromaffin cells (*b*), activity that ordinarily tends to maintain or raise blood pres- sure. When tyrosine is administered, it affects the rapidly firing brain- stem neurons; those neurons make more norepinephrine, their inhibi- tory effect is augmented (*c*) and the blood pressure falls. In response to low blood pressure, on the other hand, the firing of the brain-stem neurons is suppressed (*d*); as a result their inhibitory effect is reduced and the activity of the ganglia and the adrenal cells is increased (*e*). Now it is the high-activity sympathetic and chromaffin cells that are sensitive to increased tyrosine when the amino acid is adminis- tered. Their activity is further enhanced (*f*) and blood pressure rises.

answer may determine whether these compounds are adequately tested for therapeutic efficacy and safety and, if they pass such tests, whether they be- come available for clinical applications. Calling them drugs might lead regulato- ry agencies to require preliminary tests so laborious and expensive that they frighten away some of the most like- ly developers: the food companies that produce lecithin and protein constitu- ents for inclusion in foods.

Clearly the prescription of these sub- stances by physicians (or their recom- mendation for use in self-medication) must await the accumulation of evi- dence that they are safe and effective. Need the evidence be of the same weight for a ubiquitous food constituent as the evidence legitimately required for a drug? It probably need not be, for a number of reasons. Amino acids and choline are rapidly metabolized in the body by enzymes that have been doing the job for as long as animals have exist- ed. How an individual molecule of one

of these nutrients is metabolized seems to be largely unrelated to whether it is ingested as the pure substance or as a constituent of a food protein or leci- thin. The precursors are water-soluble and are rapidly exchanged between the bloodstream and various tissues by dif- fusion (or, in the brain, by a transport system); their tissue concentration, un- like the concentration of many drugs, does not continue to rise with repeat- ed administration. Perhaps more impor- tant, the brain must acquiesce, so to speak, in most of their effects on neuro- transmission: by means of the selective response of neurons to additional pre- cursor the brain modulates the precur- sor's effects on the brain.

The amounts of these nutrients that must be administered in order to modify neurotransmitter synthesis are relative- ly large, reflecting the fact that the nutri- ents are components of a normal diet; their normal blood level and daily in- take are not zero, as is the case with a typical drug about to be given to a pa-

tient for the first time. Hence the most convenient way of administering them might often be as a constituent of a "spe- cial food for medical uses" prepared specifically for that purpose. Such ter- minology should not imply, of course, that the diseases and conditions in which they might be therapeutic are nutrition- al in origin. Tardive dyskinesia is not the result of lecithin deficiency, nor is shock the result of tyrosine deficiency. In such conditions the nutrient would be given for its pharmacological effect, as coffee, brandy, bran or prune juice can be giv- en, as if it were a drug.

In a broader sense, however, it may be correct to think in terms of nutrition. Is it possible that old people in whom the brain has lost many of the neurons that release precursor-dependent transmit- ters might benefit from a routine diet enriched in tyrosine or tryptophan or lecithin? If they might, should their con- tinued reliance on normal but unen- riched diets be construed as constituting poor nutrition? I wonder.

III

DEVELOPMENT
OF THE BRAIN

DEVELOPMENT OF THE BRAIN

III

INTRODUCTION

The human brain grows and develops from a single fertilized cell into the most complex structure in existence. The embryonic human brain's rate of growth is staggering—over the 9 months of development, it gains neurons at an average rate of 250,000 per minute! These billions of nerve cells seem at first to develop chaotically but then migrate to their preordained destinations to form the same myriad of complex circuits, pathways, and wiring diagrams that exist in all adult human brains. Indeed, the major circuits in the brain are basically the same in all mammals. There is a very high degree of predetermination, of "hard wiring" in the brain. The ultimate plan for this of course is genetic—the genes interact with the cellular environment as the single fertilized egg eventually develops into an adult.

How the human brain grows and develops is a profound mystery. People often mistakenly think that a complete "blueprint" of the person, including the brain, is present in the genetic material (DNA) or the chromosomes. Not so. The DNA provides the "prime contracts." It specifies the plan for the building blocks (proteins) and basic rules for the way the job can be done. However, a great deal of additional information is present in the cell outside the nucleus. Specialized materials in different parts of the fertilized cell provide "subcontracts" that give essential additional information. Later, as tissues develop, the interactions among the growing tissues provide still more specific subcontracts.

The key question about the growth of the brain is how the precise, detailed, and extremely complicated neuronal circuits develop. Answers to this question ultimately will be found at the level of the genes and of single neurons and their interactions, their processes of growth and migration, and the related physical/chemical events.

This section of the reader consists of a comprehensive article by W. Maxwell Cowan of the Salk Institute. He provides a very clear overview on the growth and development of the human brain and then focusses on our current understanding of the mechanisms involved.

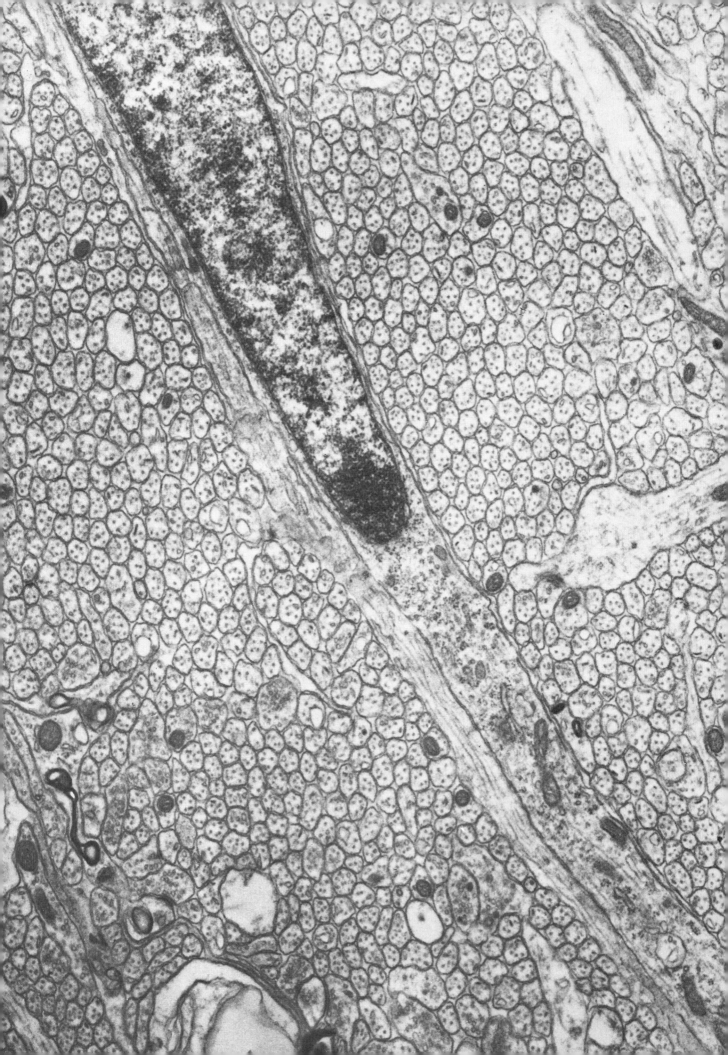

6

The Development of the Brain

by W. Maxwell Cowan
September 1979

*As the human brain develops in utero it gains neurons
at the rate of hundreds of thousands a minute. One
problem of neurobiology is how the neurons find
their place and make the right connections*

The gross changes that take place during the embryonic and fetal development of the brain have been known for almost a century, but comparatively little is known about the underlying cellular events that give rise to the particular parts of the brain and their interconnections. What is clear is that the nervous system originates as a flat sheet of cells on the dorsal surface of the developing embryo (the neural plate), that this tissue subsequently folds into an elongated hollow structure (the neural tube) and that from the head end of the tube three prominent swellings emerge, prefiguring the three main parts of the brain (the forebrain, the midbrain and the hindbrain).

It is not on these changes in the external form of the developing brain, however, that the attention of developmental neurobiologists has focused in recent years. More interesting questions intrude. How, for instance, are the various components that constitute the major parts of the nervous system generated? How do they come to occupy their definitive locations within the brain? How do the neurons and their supporting glial cells become differentiated? How do neurons in different parts of the brain establish connections with one another? In spite of a great deal of research effort it is still not possible to give a complete account of the development of any part of the brain, let alone of the brain as a whole. By determining what the main events in neural development are, however, one can begin to see how the critical issues are likely to be resolved.

Eight major stages can be identified in the development of any part of the brain. In the order of their appearance they are (1) the induction of the neural plate, (2) the localized proliferation of cells in different regions, (3) the migration of cells from the region in which they are generated to the places where they finally reside, (4) the aggregation of cells to form identifiable parts of the brain, (5) the differentiation of the immature neurons, (6) the formation of connections with other neurons, (7) the selective death of certain cells and (8) the elimination of some of the connections that were initially formed and the stabilization of others.

The process whereby some cells in the ectoderm, or outer layer, of the developing embryo become transformed into the specialized tissue from which the brain and spinal cord develop is called neural induction. It has been known since the 1920's that the critical event in neural induction is an interaction of the ectoderm and a part of the underlying layer of tissue called the mesoderm. The nature of this interaction remains to be elucidated, but there are good reasons for thinking that it involves the specific transfer of substances from the mesoderm to the ectoderm, and that as a result of that transfer the generalized tissue of the ectoderm becomes irreversibly committed to the formation of neural tissue. It is also clear that the sequential interaction of different parts of the ectoderm and the mesoderm results in the regional deter-

mination of the major parts of the future brain and spinal cord. The first part of the mesoderm to become associated with the ectoderm specifically induces forebrain structures, the next part leads to the formation of midbrain and hindbrain structures, and the last part to grow under the ectoderm is responsible for the later formation of the spinal cord.

Exactly how these regional determinations are brought about remains baffling. Experiments with disaggregated ectodermal and mesodermal cells from embryos of the appropriate age suggest that the critical element may be the relative concentration of two factors that are thought to be proteins with a low molecular weight. One of these, the neuralizing factor, seems to "prime" the ectoderm and to ensure its future neural character; the other, the mesodermalizing factor, appears in differing concentrations to determine the regional differences within the ectoderm.

Although a major effort was made in the 1930's and 1940's to isolate the putative inducing agents, it is clear in retrospect that much of the work was premature. Only in the past two decades has anything substantial been learned about the nature of gene induction generally, and it is still far from evident that the inductive mechanisms that have been identified in microorganisms operate in the same way in animal cells. There is another reason the problem of neural induction has proved to be so intractable. The only assay system suitable for the study of neural induction is ectoderm taken from embryos of the appropriate age, and since there is a limited period in development when the ectoderm is able to respond to the relevant inductive signals, it is necessary to work with extremely small amounts of tissue. Indeed, it is a tribute to the ingenuity and experimental skill of those who have addressed this problem that so much progress has already been made.

Once the major regions of the developing nervous system have been determined their potentialities become progressively limited as development pro-

MIGRATION OF A YOUNG NEURON from its birthplace deep in the cerebellum of a fetal monkey toward its final destination closer to the outer surface of the developing brain is captured in the electron micrograph on the opposite page, made by Pasko Rakic of the Yale University School of Medicine. The migrating neuron is the broader of the two diagonal bands running all the way across the micrograph from the top left to the bottom right; the dark, oblong object inside the upper part of this band is the nucleus of the nerve cell. The lighter, narrower band along the underside of the neuron is the elongated process of a glial cell, which serves both as a supporting structure and as a guide for the migrating neuron. The neuron travels through a dense neuropile, or feltwork of nerve fibers, which run in various directions. (Most of the circular structures in this view, for example, are the cross sections of axons that run more or less at right angles to the plane of the page.) Although the migrating neuron is therefore in contact with thousands of other cellular processes, it remains intimately associated with the glial cell along its entire length. This section of brain tissue is magnified about 25,000 diameters.

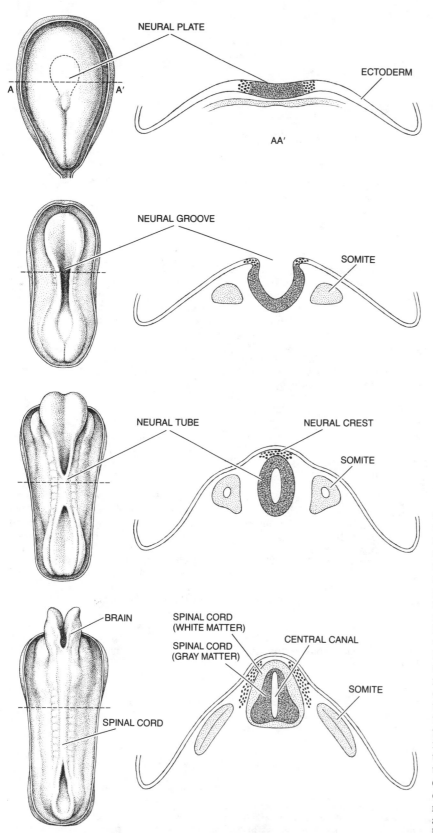

NEURAL PLATE

ECTODERM

AA'

NEURAL GROOVE

SOMITE

NEURAL TUBE

NEURAL CREST

SOMITE

BRAIN

SPINAL CORD
(WHITE MATTER)

SPINAL CORD
(GRAY MATTER)

CENTRAL CANAL

SOMITE

SPINAL CORD

GENESIS OF THE NERVOUS SYSTEM from the ectoderm, or outer cell layer, of a human embryo during the third and fourth weeks after conception is represented in these four pairs of drawings, which show both an external view of the developing embryo (*left*) and a corresponding cross-sectional view at about the middle of the future spinal cord (*right*). The central nervous system begins as the neural plate, a flat sheet of ectodermal cells on the dorsal surface of the embryo. The plate subsequently folds into a hollow structure called the neural tube. The head end of the central canal widens to form the ventricles, or cavities, of the brain. The peripheral nervous system is derived largely from the cells of the neural crest and from motor-nerve fibers that leave the lower part of the brain at each segment of the future spinal cord.

ceeds. For example, the entire head end of the neural plate initially constitutes a forebrain-eye field from which both the forebrain and the neural part of the eye develop. If one removes a small piece of ectodermal tissue at this stage, the defect is quickly replaced by the proliferation of the neighboring cells, and the development of both the forebrain and the eye proceeds quite normally. If the same operation is done at a slightly later stage, there is a permanent defect in either the forebrain or the eye, depending on the location of the piece of tissue that was removed. In other words, at this later stage it is possible to identify a forebrain field that will give rise to definitive forebrain structures, and an eye field that will form only the neural part of the eye.

At still later stages specific regions of the forebrain become delimited within the overall forebrain field. With the aid of a variety of cell-marking techniques it has been possible to construct "fate maps" that define rather precisely the final distribution of the cells in each part of the early forebrain field [*see illustration on page 70*]. The factors that lead to this progressive blocking out of smaller and smaller units, giving rise to specific parts of the brain, are not known, but it is not unreasonable to suppose that when more is learned about cellular differentiation in general, the problem will be clarified.

From studies of the embryos of amphibians it appears that the number of cells in the neural plate is comparatively small (on the order of 125,000) and that this number does not change much during the formation of the neural tube. Once the neural tube has been closed off, however, cell proliferation proceeds at a brisk pace, and before long the simple layer of epithelial cells that formed the neural plate is transformed into a rather thick epithelial layer in which the cell nuclei reside at several levels. Microscopic examination of the cells, aided in some cases by the use of radioactively labeled thymidine, a specific DNA precursor, has established that all the cells in the wall of the neural tube are capable of proliferation and that the characteristic "pseudostratified" appearance of the epithelium is attributable to the fact that the nuclei of the cells are at different levels. The nuclei synthesize DNA while they lie in the depths of the epithelium, and then they migrate toward the ventricular surface and withdraw their peripheral processes before dividing. After mitosis (cell division) the daughter cells re-form their peripheral processes, and their nuclei return to the deeper part of the epithelium before reentering the mitotic cycle. The migration of the nuclei of proliferating neurons is characteristic of epithelial cells of this kind.

After the cells pass through a number

of such cycles (the number varies from region to region and from population to population within any one region) they apparently lose their capacity for synthesizing DNA, and they migrate out of the epithelium to form a second cellular layer adjacent to the ventricular zone. The cells that constitute this mantle, or intermediate, layer are young neurons, which never again divide, and glial-cell precursors, which retain their capacity for proliferation throughout life.

Although it is not known what turns the proliferative mechanism on and off in any region of the nervous system, it is clear that the relative times at which different populations of cells cease dividing is rigidly determined, and there is now a sizable body of evidence to suggest that this is a critical stage in the life of all neurons. Not only does the withdrawal of a cell from the mitotic cycle seem to trigger its subsequent migration into the intermediate layer but also the

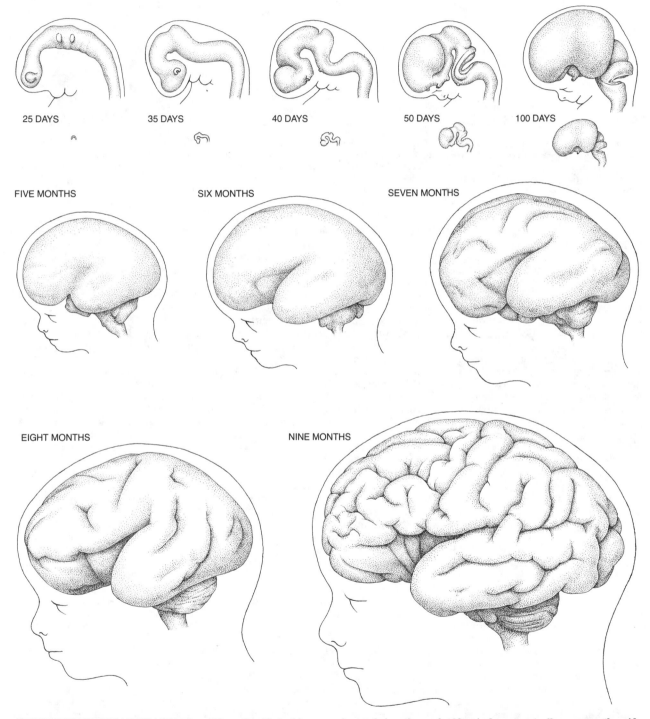

25 DAYS   35 DAYS   40 DAYS   50 DAYS   100 DAYS

FIVE MONTHS   SIX MONTHS   SEVEN MONTHS

EIGHT MONTHS   NINE MONTHS

DEVELOPING HUMAN BRAIN is viewed from the side in this sequence of drawings, which show a succession of embryonic and fetal stages. The drawings in the main sequence (*bottom*) are all reproduced at the same scale: approximately four-fifths life-size. The first five embryonic stages are also shown enlarged to an arbitrary common size to clarify their structural details (*top*). The three main parts of the brain (the forebrain, the midbrain and the hindbrain) originate as prominent swellings at the head end of the early neural tube. In human beings the cerebral hemispheres eventually overgrow the midbrain and the hindbrain and also partly obscure the cerebellum. The characteristic convolutions and invaginations of the brain's surface do not begin to appear until about the middle of pregnancy. Assuming that the fully developed human brain contains on the order of 100 billion neurons and that virtually no new neurons are added after birth, it can be calculated that neurons must be generated in the developing brain at an average rate of more than 250,000 per minute.

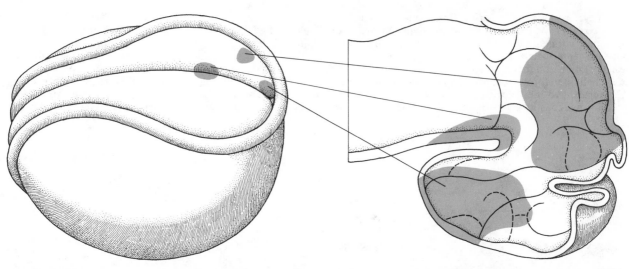

DERIVATION of each of the major regions of the brain can be traced by labeling different regions of the neural plate of an experimental animal at a very early embryonic stage with the aid of a variety of cell-marking techniques. In this demonstration of how such "fate maps" are constructed three regions have been marked on the neural plate of the early embryo of the axolotl, a large amphibian (*left*). The final positions of the cells in the marked regions are then plotted as they would be seen in a sagittal section of the brain at a later stage of embryonic development (*right*). This illustration is adapted from the work of D. C.-O. Jacobson of the University of Uppsala.

cell seems at the same time to acquire a definitive "address," in the sense that if its "birth date" (defined as the time when a cell loses its capacity for DNA synthesis) is known, it is possible to predict where the cell will finally reside. Furthermore, it seems in some cases that the pattern of connections the neuron will ultimately form is also determined at this time.

From experiments in which small amounts of radioactively labeled thymidine have been administered to embryos (or in the case of mammals to their pregnant mothers) investigators now know the birth dates of the cells in many parts of the brain for a number of different species. From these studies it is now possible to make several generalizations about the patterning of cell proliferation in the brain. First, the larger neurons, including most of the cells whose processes extend for considerable distances, such as the cells in the retina that project to the visual centers of the brain, are usually generated earlier than the smaller neurons, whose fibers are confined to the region of the cell body. Second, the sequence of cell proliferation is characteristic for each region of the brain. For example, in the cerebral cortex the first cells to withdraw from the proliferative cycle will in time come to occupy the deepest cortical layer, whereas those that are generated at successively later times form the progressively more superficial layers of the cortex.

On the other hand, in the neural retina (which is actually an extension of the brain) the sequence of cell proliferation is essentially the reverse; the first population of cells to be generated (the ganglion cells) migrates to the most superficial layer of the retina, and subsequent populations of cells occupy progressively deeper layers. In other regions of the

brain the sequences are more complex, but in each region it is evident that cells occupying similar positions are always generated at the same time; conversely, cells generated at different times invariably come to reside in different zones within the region. A third generalization that can be made is that in most parts of the brain the first supporting cells to be formed appear at about the same time as the first neurons, but as a rule the proliferation of glial cells continues for a much longer period.

The number of neurons initially formed in any region of the brain is determined by three factors. The first factor is the duration of the proliferative period as a whole; in the regions that have been studied to date, it has been found to range from a few days to several weeks. The second factor is the duration of the cell cycle; in young embryos it is usually on the order of a few hours, but as development progresses it may become as long as four or five days. The third factor is the number of precursor cells from which the neuronal population is derived.

A number of methods are now available for determining the duration of the proliferative period and the length of the cell cycle, but except in a few cases it is not possible to estimate the size of the precursor pool of cells. Part of the reason for this difficulty is that it is impossible at present to follow the fate of individual cells in the developing mammalian brain, as has been done in the much simpler nervous systems of certain invertebrates. In these organisms the embryos are often quite transparent, and individual cells can be followed through several mitotic divisions with the aid of a light microscope equipped with differential-interference optics. Alternatively, the precursor cells in such organisms

may be so large that they can be readily labeled by the intracellular injection of marker molecules such as horseradish peroxidase; if the marker is not degraded, it can be distributed to all the cell's progeny, at least over several cell generations.

Since most neurons are generated in or close to the ventricular lining of the neural tube and finally come to rest at some distance from this layer, they have to go through at least one phase of migration after withdrawing from the proliferative cycle. There are a few situations in which cells migrate away from the ventricular zone but continue to proliferate. This is usually observed in a special region found between the ventricular and the intermediate zone, known as the subventricular zone. This layer, which is particularly prominent in the forebrain, gives rise to many of the smaller neurons in some of the deep structures of the cerebral hemisphere (the basal ganglia), to certain small cortical neurons and to many of the glial cells in the cerebral cortex and the underlying white matter. In the hindbrain some of the cells in the corresponding subventricular region undergo a second migration under the surface of the developing cerebellum, where they set up a special proliferative zone known as the external granular layer. In the human brain proliferation in this layer continues for several weeks and gives rise to most of the interneurons in the cerebellar cortex, including the billions of granule cells that are a distinctive feature of the cerebellum. With these and a few other exceptions, most migrations of neurons involve the movement of postmitotic cells.

The process of neuronal migration appears in most cases to be amoeboid.

The migrating cells extend a leading process that attaches itself to some appropriate substrate; the nucleus flows or is drawn into the process, and the trailing process behind the nucleus is then withdrawn. It is a fairly slow procedure, the average rate of migration being on the order of a tenth of a millimeter per day. In a few cases the cell as a whole does not migrate. Instead it begins to form some of its processes at an early stage in its development, and later the cell body begins to move progressively farther away from the first processes, which remain essentially where they originated.

Since neurons often migrate over considerable distances, it would be interesting to know to what types of directional cue they respond. In particular, how do they know when to stop migrating and to begin aggregating with other neurons of the same kind? It has been known for some time that there are specialized glial cells within the developing brain whose cell bodies lie inside the ventricular zone and whose processes extend radially to the surface. Since these cells appear at an early stage of development and persist until some time after neuronal migration has ceased, it has been suggested that they might provide an appropriate scaffolding along which the migrating neurons might move. Certainly in electron micrographs of most parts of the developing brain the migrating cells are almost invariably found to be closely associated with the neighboring glial processes. This relation has led Pasko Rakic of the Yale University School of Medicine to postulate that migrating cells are directed to their definitive locations by such glial processes. In support of this view Rakic and Richard L. Sidman of the Children's Hospital Medical Center in Boston have noted that in one of the most striking genetic mutations affecting the cerebellum of the mouse the radial glial processes degenerate at a comparatively early stage; apparently as a result of this degeneration the migration of most of the granule cells is severely disrupted.

Considering the distances over which many neurons move in the course of development, it is perhaps not surprising that during their migration some cells are misdirected and end up in distinctly abnormal positions. Such neuronal misplacements (termed ectopias) have long been recognized by pathologists as a concomitant of certain gross disorders in brain development, but it is not generally appreciated that even during normal development a proportion of the migrating cells may respond inappropriately to the usual directional cues and end up in aberrant locations. Recent technical advances have made it possible to recognize cells of this kind in several situations, and it is significant that the majority of such misplaced neurons appear to be eliminated during the later

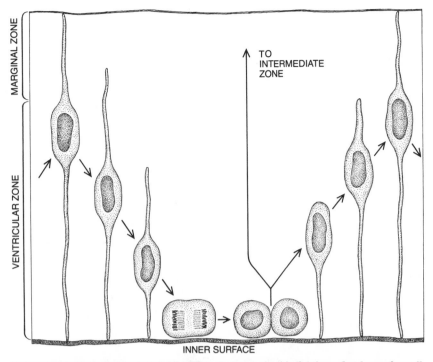

NUCLEI OF NERVE CELLS MIGRATE in the layer of epithelial tissue that forms the wall of the neural tube in the developing embryo, as this multistage schematic diagram shows. When the cells in this layer, called the neuroepithelium or ventricular zone, replicate their DNA, their nuclei migrate toward the inner surface of the epithelium, their peripheral processes become detached from the outermost layer and the cells become rounded before dividing. After mitosis (cell division) the daughter cells either extend a new process so that their nuclei can migrate back to the middle level of the epithelium, or (if the cells have stopped dividing) they migrate out of the epithelium to form part of the intermediate zone in the wall of the brain.

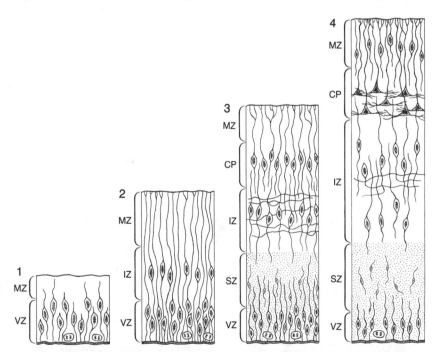

PROGRESSIVE THICKENING of the wall of the developing brain is illustrated. At the earliest stage (1) the wall consists only of a "pseudostratified" epithelium, in which the ventricular zone (VZ) contains the cell bodies and the marginal zone (MZ) contains only the extended outer cell processes. When some of the cells lose their capacity for synthesizing DNA and withdraw from the mitotic cycle (2), they form a second layer, the intermediate zone (IZ). In the forebrain the cells that pass through this zone aggregate to form the cortical plate (CP), the region in which the various layers of the cerebral cortex develop (3). At the latest stage (4) the original ventricular zone remains as the ependymal lining of the cerebral ventricles, and the comparatively cell-free region between this lining and the cortex becomes the subcortical white matter, through which nerve fibers enter and leave the cortex. Subventricular zone (SZ) is a second proliferative region in which many glial cells and some neurons in forebrain are generated.

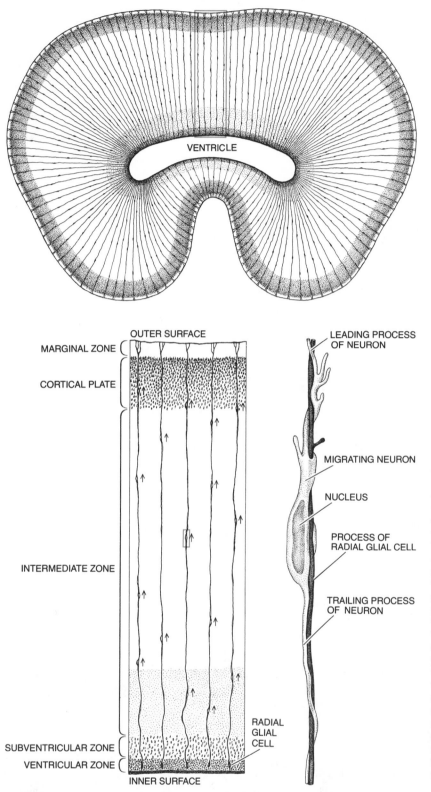

stages of development. In one population of neurons that has been carefully studied from this point of view, about 3 percent of the cells have been found to migrate to some abnormal location; all but a handful of these misplaced neurons, however, degenerate during the later phase of naturally occurring cell death.

When the migrating neurons reach their definitive locations, they generally aggregate with other cells of a similar kind to form either cortical layers or nuclear masses. The tendency of developing cells of the same embryonic origin to selectively adhere to one another was first demonstrated more than 50 years ago, but it is only in the past decade that this subject has attracted the attention it deserves from neuroembryologists. Much of the initial stimulus for the more recent work stemmed from the search for the molecular mechanisms underlying the formation of specific connections between related groups of neurons. Unfortunately that problem has proved to be intractable, but much of the work that was done on it bears directly on the important issue of how discrete populations of neurons are formed in the developing brain.

Perhaps the most important finding to come out of these studies is that when cells from two or three regions of the developing nervous system are dissociated (usually mechanically or by mild chemical treatment), mixed together and then allowed to reaggregate in an appropriate medium, they tend to sort themselves out so that the cells from each region preferentially aggregate with other cells from the same region. This selective adhesiveness seems to be a general property of all living cells and is probably due to the appearance on their surfaces of specific classes of large molecules that serve both to "recognize" cells of the same kind and to bind the cells together. These molecules, which function as ligands between cells, appear to be highly specific for each major type of cell. Moreover, they appear to change in either number or distribution as development proceeds. At present workers in several laboratories are endeavoring to isolate and characterize these and other surface ligands, and it seems likely that this may be the first major problem in neural development to be successfully analyzed at the molecular level.

One special feature of cell aggregation in the developing nervous system is that in most regions of the brain the cells not only adhere to one another but also adopt some preferential orientation. For example, in the cerebral cortex the majority of the large pyramidal neurons are consistently aligned with their prominent apical dendrites directed toward the surface and their axons directed toward the underlying white matter. It is not evident how cells come to be aligned

SPECIALIZED SUPPORTING CELLS, the radial glial cells, arise during the early stages of the development of the nervous system. These cells are distinguished by their extremely long processes, which span the entire thickness of the wall of the neural tube and its derivative structures. The drawing at the top shows how the radial glial cells look in a Golgi-stained preparation of a thick transverse section through the wall of the cerebral hemisphere of a fetal monkey. The cell bodies lie in the ventricular zone and their processes extend to the outer surface of the surrounding layers, where they appear to form expanded terminal attachments. An enlarged view of a segment of this transverse section is shown at the bottom left. The small portion of tissue inside the colored rectangle in this enlargement is further magnified in the detailed three-dimensional view at the bottom right, which is based on the microscopic studies of Rakic. This illustration reveals the close relation between the processes of the radial glial cells and the migrating neurons, a relation that is observed in the development of most parts of the brain.

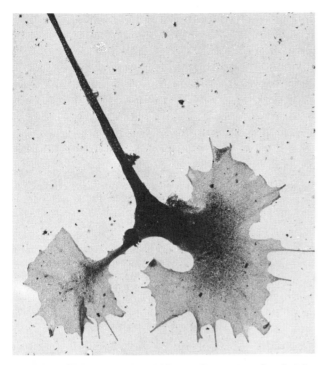
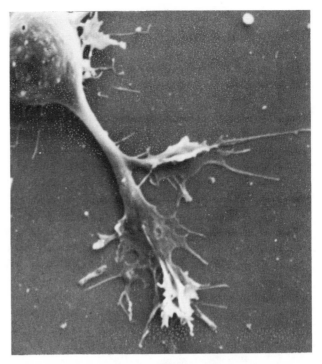

GROWTH CONES (expanded, highly motile structures found at the ends of growing neuronal processes) are seen in these two electron micrographs. The transmission electron micrograph at the left shows a pair of growth cones at the end of an axonlike process of a sympathetic ganglion cell from a rat. The cell had been dissociated and grown in tissue culture, and the process seen here had branched just a few minutes before the cell was fixed and prepared (without sectioning) for viewing in the electron microscope. The fine, fingerlike extensions are filopodia; the flattened veil-like sheets between them are lamellipodia. The scanning electron micrograph at the right shows a growing dendrite of a neuron obtained from the hippocampus of a fetal rat's brain. The growth cones in this surface view were formed after the neuron had been dissociated and grown in tissue culture for only two hours. Both pictures were made at the Washington University School of Medicine, the one at the left by J. Michael Cochran and Mary Bartlett Bunge, the one at the right by Steven R. Rothman.

in this way, but it seems likely that it is attributable either to the existence of different classes of cell-surface molecules specifically concerned with cell-cell orientation or to the selective redistribution of surface molecules responsible for the cell's initial aggregation.

One of the most striking features in the development of neurons is the progressive elaboration of their processes, but this is only one aspect of their differentiation. Equally important are their adoption of a particular mode of transmission (most neurons generate action potentials but some show only decremental transmission) and the selection of one or the other of two modes of interaction with other cells (either by the formation of conventional synapses to provide for the release of a chemical transmitter or by the formation of gap junctions to provide for electrical interactions of cells). Neurobiologists are only now beginning to learn something of these more covert aspects of neuronal differentiation, and it is becoming clear that neurons may be considerably more complex than had been imagined. For example, it has recently been shown that some neurons can switch from one chemical transmitter to another (specifically from norepinephrine to acetylcholine) under the influence of certain environmental factors, whereas others can show a change in the principal ion they use for the propagation of nerve impulses at different developmental stages (changing, for instance, from calcium to sodium).

Rather more is known about the formation of neuronal processes. Most neurons in the brain of mammals are multipolar, with several tapering dendrites, which generally function as receptive processes, and a single axon, which serves as the cell's main effector process. Although some cells are known to form processes before they start migrating, the majority begin to generate processes only after reaching their final position. Exactly what stimulates the formation of processes is not clear. Studies in which immature neurons have been isolated and maintained in tissue culture reveal that processes are formed only when the cells are able to adhere to an appropriate substrate and that under these conditions the cells are often able to form a fairly normal complement of both dendrites and axons. In some cases, in spite of the highly artificial conditions in which the neurons are grown, the overall appearance of the dendrites that are formed closely resembles that seen in the intact brain, even though the cells are deprived of all contact with other neurons or even glial cells. Observations of this kind suggest that the information required for a neuron to generate its distinctive dendritic branching is genetically determined.

It is also evident, however, that during the normal development of the brain most neurons are subject to a variety of local mechanical influences that may modify their form. Certainly the number and distribution of the inputs the cells receive may critically affect their final shape. A striking example of this effect is seen in the cerebellum. The dendrites of the most distinctive class of neurons in the cerebellar cortex, the Purkinje cells, normally have a characteristic planar arrangement that is oriented at right angles to the axons of the granule cells that constitute their principal input; if for any reason the usual regular arrangement of the granule cells' axons is disrupted, the planar distribution of the dendrites of the Purkinje cells is correspondingly altered.

The actual mechanism by which the processes of a neuron are elongated is now quite well understood. Most processes bear distinctive structures at their growing ends called growth cones. These expanded, highly motile structures, which in the living state seem to be continually exploring their immediate environment, are the sites where most new material is added to the growing process. When a process branches, it almost always does so by the formation of a new growth cone. Although the evidence is largely indirect, there are reasons for thinking the growth cone has encoded within (or on) it the necessary

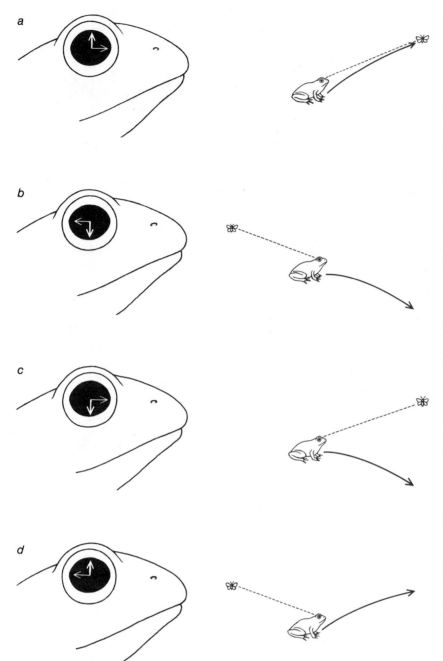

ONE EXPERIMENTAL APPROACH to the study of how neurons make specific patterns of connections in the developing brain involves manipulating the projection of the retina on the optic tectum of the midbrain. In this approach, pioneered by Roger W. Sperry of the California Institute of Technology, the eyes of adult frogs (or tadpoles at various stages in development) are rotated or transplanted. Later, when the optic nerve has regenerated (or when the tadpoles have grown into frogs and the axons of the retinal ganglion cells, which comprise the optic nerve, have made connections in the optic tectum), one can observe what effect the operation has had on the frogs' visual behavior; one can also map the retinal projection on the tectum electrophysiologically. This series of drawings, based on the work of Sperry, shows first the behavior of a control frog with its eyes in the normal positions (a). In experiment b the right eye had been rotated through 180 degrees; when the frog was tested some time after the optic nerve had regenerated, the frog's attempt to strike at a lure placed in its upper field of view was exactly 180 degrees in the wrong direction. In experiment c the left eye had been substituted for the right eye, inverting only the dorsoventral axis (*thick arrow*); in this case the frog directed its strike forward toward the lure, but in the direction of the lower visual field instead of the upper one. In experiment d a similar transplantation had been done, but this time inverting the eye only in the anteroposterior direction (*thin arrow*); the frog sensed that the lure was in the upper visual field, but it now struck forward instead of backward. The outcome of these experiments is consistent with the view that during regeneration the fibers of the optic nerve always grow back to the part of the optic tectum they originally innervated, and that during normal development they similarly "find their way" to their correct positions in the tectum. These findings are best explained by the hypothesis that both the retinal ganglion cells and their target neurons in the tectum have chemical characteristics that enable them to identify each other.

molecular features that enable it both to detect appropriate substrates along which to grow and to identify appropriate targets. Experiments in which neurons have been grown on a variety of artificial substrates indicate that most processes grow preferentially along surfaces of high adhesiveness.

One of the least well understood problems in the entire field of developmental neurobiology is how axons are able to find their way. It is particularly difficult to see how they do so when they may have to extend for considerable distances within the brain and at one or more points along their course deviate either to the right or to the left, cross to the opposite side of the brain and give off one or more branches before finally reaching their predetermined destination. In some systems it looks as if the axons simply grow under the influence of certain gradients that act along the major axes of the brain and spinal cord; in other systems the axons seem to be guided by their relation to their nearest neighbors. In many cases, however, it appears that the growing axon has encoded within it a sophisticated molecular mechanism that enables it to correctly respond to structural or chemical cues along its route.

Such directionally guided growth has recently been demonstrated by Rita Levi-Montalcini of the Laboratory of Cell Biology of the National Research Council in Rome. When she and her colleagues injected the protein known as nerve-growth factor into the brain of young rats, there was an abnormal growth of axons from sympathetic ganglion cells (peripheral neurons that lie alongside the vertebral column and are known to be sensitive to nerve-growth factor) into the spinal cord and up toward the brain, apparently along the route of diffusion of the injected nerve-growth factor. In this case the nerve-growth factor was acting not so much as a trophic, or growth-promoting, substance (as it usually does) but rather as a tropic, or direction-determining, substance, and the sympathetic nerve axons were responding chemotropically to its presence.

There are two other features of the growth of nerve processes that merit comment. The first is that most neurons seem to generate many more processes than are needed or than they are subsequently able to maintain. Hence most young neurons bear large numbers of short dendritelike processes, all but a few of which are later retracted as the cells mature. Similarly, most developing axons appear to make many more connections than are needed in the mature state, and commonly there is a phase of process elimination during which many (and in some cases all but one) of the initial group of connections are withdrawn. The second feature is that there is a strong tendency for axons to grow in close association with their neighbors, a phenomenon known as fasciculation. Recent work suggests that the tendency to fasciculate may be associated with the appearance along the length of most axons of surface ligands that enable them to join up and grow with other axons of a similar kind. In at least one instance it seems that because of this type of lateral association only the first axon in the group needs to develop a conventional growth cone; the other axons simply follow the leader.

Undoubtedly the most important unresolved issue in the development of the brain is the question of how neurons make specific patterns of connections. Earlier notions that most of the connectivity of the brain was functionally selected from a randomly generated set of connections are now seen to be untenable. Most of the connections seem to be precisely established at an early stage of development, and there is much evidence that the connections formed are specific not only for particular regions of the brain but also for particular neurons (and in some cases particular parts of the neurons) within these regions.

Several hypotheses have been put forward to explain how this remarkable precision is brought about. Some workers have argued that it can be simply explained on the basis that growing axons maintain the same topographical relation to one another as their parent cell bodies have. Others have suggested that the timing of events (in particular the time at which different groups of fibers reach their target regions) is critical. The one explanation that seems to fit all the observed phenomena is the chemoaffinity hypothesis, first formulated by Roger W. Sperry of the California Institute of Technology. According to this view most neurons (or more likely most small populations of neurons) become chemically differentiated at an early stage in their development depending on the positions they occupy, and this aspect of their differentiation is expressed in the form of distinguishing labels that enable the axons of the neurons to recognize either a matching label or a complementary one on the surface of their target neurons.

Although the problem is a general one affecting all parts of the nervous system, it has been most intensively studied in two systems: the innervation of the limb musculature by the relevant motor neurons in the spinal cord and the projection of the ganglion cells in the retina of the eye to their principal terminus in the brain of lower vertebrates, the optic tectum. Studies of muscle innervation indicate that under normal circumstances small populations of motor neurons, called motor-neuron pools, become segregated at an early stage in development, and that each motor-neuron pool

preferentially innervates a specific limb muscle, few errors being made in the process. Although the specificity of the innervation pattern is normally precise, it is not absolute. Hence if a supernumerary hindlimb from a donor chick embryo is transplanted alongside the normal hindlimb of a host embryo, the muscles in the supernumerary limb invariably become innervated by motorneuron pools that normally innervate either parts of the trunk or the limb-girdle musculature. The pattern of innervation is clearly aberrant, but the fact that the muscles in the transplanted limb are always innervated by the same populations of cells strongly suggests that even under these unusual conditions the axons of the motor neurons obey some (as yet unidentified) set of rules.

The retinotectal system has proved to be particularly advantageous for the analysis of the problem. In amphibians it is possible at the embryonic and larval stages to carry out a variety of experimental manipulations such as rotating the eye, making compound eyes with pieces of tissue obtained from different segments of two or more retinas and ablating or rotating parts of the tectum. Later, when the system is fully developed, it is quite easy to determine the connections formed by the retinal ganglion cells anatomically, electrophysiologically or behaviorally. Furthermore, in fishes and amphibians the optic nerve (which is formed by the axons of the retinal ganglion cells) is capable of regeneration after its fibers have been interrupted, so that it is possible to carry out many of the same kinds of experimental manipulation in juvenile or adult animals. Since there is now a vast body of literature on this system, only some of the major findings can be summarized here.

Perhaps the most important findings to have emerged from this work come from two main groups of experiments. In the first group of experiments an optic nerve was cut in frogs and salamanders, and the eye was rotated through 180 degrees. In the other experiments portions of the optic tectum of goldfish and frogs were excised and the excised portions were either rotated or transferred to another part of the tectum. In both groups of experiments the regenerating fibers of the optic nerve could be shown, either electrophysiologically or behaviorally, to have grown back to the same parts of the tectum as those they originally innervated. The simplest explanation for this finding is that the axons of the ganglion cells and their target neurons in the optic tectum are labeled in some way, and that the regenerating axons grow back until they "recognize" the appropriate labels on the neurons in the relevant part of the optic tectum.

It is difficult to refute the argument

that under such circumstances the fibers from different parts of the retina had earlier "imprinted" themselves on the related groups of tectal cells, and that the axons or the tectal neurons simply "remembered" their previous position. There is some evidence to suggest, however, that a similar mechanism may account for the initial development of the system. If the developing eye of a frog is rotated before a certain critical stage in development, the resulting projection of the retina on the tectum tends to be normal. If the rotation is done after the critical period, however, the retinal projection is invariably rotated to the same degree. Similarly, if the entire embryonic optic tectum is rotated by 180 degrees in the head-to-tail dimension (together with a portion of the forebrain that lies just in front of it), the retinal projection that is formed is again inverted.

These experiments suggest that there is a certain stage in the development of most neural centers during which they become topographically polarized in such a way that the constituent neurons acquire some determining characteristic that establishes the spatial organization of the projection as a whole. Marcus Jacobson of the University of Miami School of Medicine showed some years ago that in the clawed frog Xenopus laevis the retina becomes polarized in this way at about the time the first ganglion cells withdraw from the mitotic cycle. Although at this stage only about 1 percent of the ganglion cells are present, the entire future patterning of the retinal projection on the tectum seems to be established at the same time. It is not at all clear how neurons acquire positional information of this type or how it is expressed in the outgrowth of their processes. It appears, however, that the polarity-determining mechanisms are not confined to the nervous system but operate throughout the organism. R. Kevin Hunt of Johns Hopkins University and Jacobson have found that if a developing eye is transplanted into the flank of a larval frog before the period of axial specification and allowed to pass through the critical period in this abnormal position, then when it is retransplanted into the orbit, or eye socket, the ganglion cells form connections within the optic tectum that reflect the orientation of the eye during the period it was in the flank, rather than its position after it was replaced in the orbit.

When a growing axon reaches its appropriate target, whether it is another group of neurons or an effector tissue such as a collection of muscle or gland cells, it forms specialized functional contacts—synapses—with these cells. It is at such sites that information is transmitted from one cell to another, usually through the release of small quantities of an appropriate transmitter [see "The Chemistry of the Brain," by Leslie L. Iversen, page 30]. A large body of phe-

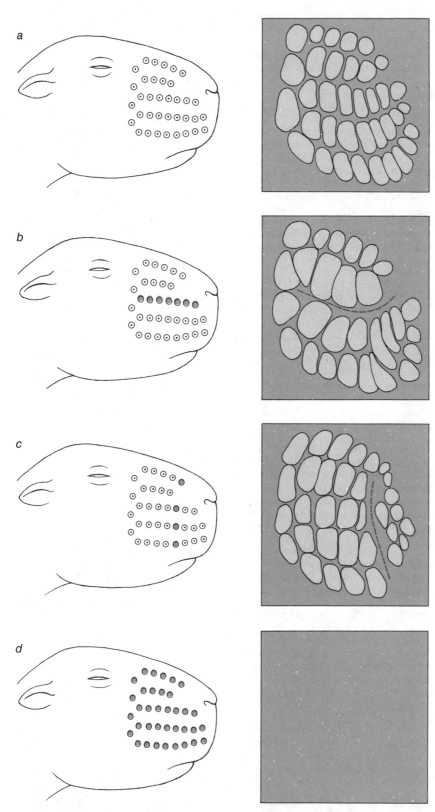

WHISKERS AND BARRELS in a young mouse are one of many systems that have been found to demonstrate the critical dependence of the developing nervous system on its inputs. The whiskers in this case are the sensory hairs on a mouse's snout; the barrels are specialized aggregations of neurons in the fourth layer of the mouse's cerebral cortex. Each barrel receives its input from a single whisker on the opposite side of the mouse's snout (a). If one row of whiskers is destroyed shortly after birth, the corresponding row of barrels in the cerebral cortex will later be found to be missing and the adjoining barrels to be enlarged (b, c). If all the whiskers are destroyed, the entire group of barrels will have disappeared (d). There must be a considerable degree of plasticity in the developing cortex, since the fibers that innervate the whiskers are not directly connected to the cortex but are linked to it through at least two synaptic relays. Illustration is based on work of Thomas A. Woolsey of Washington University School of Medicine.

nomenological evidence suggests that at synapses there is an important two-way transfer of substances essential for the survival and normal functioning of both the presynaptic and the postsynaptic cells. These substances, which are collectively referred to as trophic factors, are for the most part hypothetical; only one (nerve-growth factor) has been identified and chemically characterized. This substance, which was first identified by Viktor Hamburger and Levi-Montalcini at Washington University in the 1950's, has been found to be a protein that normally exists in the form of a pair of identical amino acid chains, each with a molecular weight of slightly more than 13,000 daltons.

Although the mode of action of nerve-growth factor has not yet been defined, it is known to be essential for the growth and survival of sympathetic ganglion cells, and during development it specifically promotes the outgrowth of processes from these and from certain spinal ganglion cells. In addition, as I have already noted, in some cases it may influence the directed outgrowth of sympathetic nerve fibers. Conversely, if an antibody to nerve-growth factor is administered to newborn mice, it leads to the destruction of the entire sympathetic nervous system. Even in adult animals nerve-growth factor appears to be continually supplied to sympathetic neurons by their target tissues, the protein being taken up by the terminal portions of their axons and transported back to the cell body. If the supply is interrupted by cutting the axons of the sympathetic neurons, their functional integrity is seriously disturbed and the synapses that end on the cells are promptly withdrawn. It seems probable that in the next few years several other substances of this kind will be isolated, and it may well be shown that most classes of neurons are dependent on a specific agent for their survival and for the directed growth of their processes.

It has become evident in recent years that the development of many structures and tissues is sculptured by highly programmed phases of cell death. This is true also of the developing brain. In many regions of the brain the number of neurons originally generated greatly exceeds the number of neurons that survive beyond the developmental period. In each region for which quantitative data are available it has been found that the number of neurons is adjusted during a phase of selective cell death that always occupies a predictable period (usually at about the time when the population of neurons as a whole is forming synaptic connections with its target tissue). It is not known whether this phenomenon operates in every part of the brain (it has been studied mainly in small groups of cells), but in those where it has been documented it involves between 15 and 85 percent of the initial neuronal population.

It seems, therefore, that in many parts of the brain the final size of the neuronal population is established in two stages: an early stage in which a comparatively large number of cells are generated, and a later stage in which the number of neurons is adjusted to match the size of the field they innervate. It is commonly assumed that the limiting factor determining the final number of cells is the number of functional contacts available to the axons of the developing neurons. Certainly if one experimentally reduces the size of the projection field, the magnitude of the naturally occurring cell death is accentuated to a proportional degree. In the case of the spinal motor neurons that innervate the hindlimb musculature it has been possible to reduce the amount of cell death in chick embryos by experimentally adding a supernumerary limb. Recent experiments suggest, however, that it may not be the formation of connections that is critical but rather the amount of trophic material available to the cells.

At a somewhat later stage in development there is a second adjustment, not in the size of the neuronal population as a whole but in the number of processes its cells maintain. The phenomenon of process (and synapse) elimination was first observed in the innervation of the limb muscles in young rats. Whereas in mature animals most muscle cells are innervated by a single axon, during the first postnatal week as many as five or six separate axons can be shown to form synapses with each muscle fiber. Over the course of the next two or three weeks the additional axons are successively eliminated, until only one axon survives. A comparable phase of process elimination has also been found in certain neuron-to-neuron connections both in the peripheral nervous system and in the brain. To cite just one example, in the cerebellum of adult animals each Purkinje cell receives only a single incoming nerve fiber of the class known as climbing fibers, but during the immediate postnatal period several such fibers may contact each Purkinje cell. Except in certain genetic mutations that affect the cerebellum all but one of these fibers are eliminated.

The finding that many early processes are later eliminated raises an interesting question: What determines which processes survive and which are eliminated? At present one can only surmise that during development fibers compete among themselves in some way. There is evidence to suggest that one factor that may give some fibers a competitive edge over the others is their functional activity. Certainly in many systems the final form of the relevant neuronal populations emerges only gradually from a rather amorphous structure, and it is often possible to alter markedly the final

appearance of the structure and its connections by interfering with its function during certain critical periods in its development. Two examples drawn from the sensory areas of the cerebral cortex will serve to make this point.

In the macaque monkey information from the retina reaches the fourth layer of the visual cortex by way of a structure called the lateral geniculate nucleus. At this level in the cortex the inputs from the two eyes are quite separate, a fact that has been directly demonstrated in experimental animals by injecting large amounts of a radioactively labeled amino acid into one eye. The retinal ganglion cells take up the labeled amino acid, incorporate it into protein and transport it to the lateral geniculate nucleus. Here some fraction of the label is released and becomes available for incorporation by the geniculate cells, which can then transport it along their axons to the visual cortex. In suitably prepared autoradiographs (in which the distribution of the labeled fibers reaching the cortex can be visualized) it is evident that the primary visual area is arranged into alternating eye-dominance bands, each band about 400 micrometers wide, that receive their input from either the right eye or the left eye. David H. Hubel, Torsten N. Wiesel and Simon LeVay of the Harvard Medical School have shown that if the eyelids of one eye of an experimental animal are sutured shut shortly after birth (so that the retina of the eye is never exposed to patterned illumination), the eye-dominance bands connected to the deprived eye are much narrower than normal bands. At the same time the bands connected with the open eye are correspondingly wider (the overall width of two adjoining bands remaining constant).

This result appears to be brought about partly by the shrinkage of the eye-dominance bands connected to the deprived eye, accompanied by a secondary expansion of those associated with the nondeprived, normal eye, and partly by the persistence of an earlier, more widespread distribution of the fibers from the nondeprived eye. If the inputs from the two eyes are examined at different stages in development, it can be shown that when the fibers from the lateral geniculate nucleus first reach the visual cortex, the inputs from one eye extensively overlap those from the other. It is not until about the end of the first postnatal month that the eye-dominance bands become clearly defined. In the light of this discovery (and the results of experiments in which the deprived eye is reopened and the other eye is sutured shut) it seems likely that the effect of visual deprivation is to place geniculo-cortical cells that are connected with a deprived eye at some disadvantage, so that they become less effective in competing for synaptic sites on the target cells in the fourth layer of the cortex.

In the corresponding layer in the sensory cortex of the mouse the cells are arranged in a number of distinctive groupings called barrels. Physiological studies have shown that each barrel receives its input from a single whisker on the opposite side of the mouse's snout, the whiskers being among the most important sense organs in mice. Thomas A. Woolsey of the Washington University School of Medicine, who first recognized the importance of the barrels, has found that if a small group of whiskers is removed during the first few days after birth, the corresponding group of barrels in the cortex fails to develop. This is a particularly interesting finding because there are at least two intervening groups of neurons between the sensory neurons that innervate the whiskers and the neurons that constitute the cortical barrels.

These and many other observations make it clear that the developing brain is an extremely plastic structure. Although many regions may be "hardwired," others (such as the cerebral cortex) are open to a variety of influences, both intrinsic and environmental. The ability of the brain to reorganize itself in response to external influences or to localized injury is currently one of the most active areas in neurobiological research, not only because of its obvious relevance for such phenomena as learning and memory, and its bearing on the capacity of the brain to recover after injury, but also because of what it is likely to reveal about normal brain development.

Finally, it is worth pointing out that the development of the brain, like the development of most other biological structures, is not without error. I have already indicated that errors may appear during neuronal migration. There are also several known cases in which errors are made during the formation of connections. In the visual system it has been noted by a number of workers that some optic-nerve fibers that should cross the midline in the optic chiasm grow back aberrantly to the same side of the brain. In some of these situations if one eye is removed from an experimental animal early in development, the number of aberrantly directed fibers can be considerably increased. Since such aberrant fibers are often not seen in the mature brain, it looks as if the misdirected axons (and whatever inappropriate connections they form) are eliminated at later stages in development. How they are recognized as being erroneous and how they are subsequently removed remains a puzzle. Considering the complexity of the developmental mechanisms involved, it is hardly surprising that errors are found. What is surprising is that they appear infrequently and that they are often effectively eliminated.

IV

SENSORY
PROCESSES

SENSORY PROCESSES

IV

INTRODUCTION

T he molding pressure of evolution is strikingly clear in the sensory systems. The sensitivity at which animals can detect stimuli is quite extraordinary. The human eye and ear are as sensitive as the physical limits permit. At the appropriate wavelength, for example, the human eye can detect one photon of light energy, the smallest measurable amount there is. The human ear can hear sounds that are much fainter than any physical instrument can detect.

Because there can be only a few optimum solutions to detecting a certain stimulus, evolution has produced an eye in the octopus that is similar in many ways to the human eye, even though their evolutionary histories are different. There are numerous examples of such convergent evolution in sensory receptors. In general, animals are most sensitive to stimuli affecting their adaptive behavior. Bats and porpoises echolocate; they they "see" their world from the reflected high-frequency sounds they emit. They can hear sounds that are five or six times higher in frequency than humans can. Honey bees can see ultraviolet light. Many flowers that look dull to us reflect large amounts of such light.

We know more about the visual system and visual sensation than any other aspect of brain function, thanks to the discoveries made by David Hubel, Torsten Wiesel, and others on the structural organization and functional properties of the mammalian visual brain. Such information as how neurons in the visual area of the cerebral cortex respond selectively to only certain aspects and orientations of visual stimuli is discussed widely in both textbooks (see, e.g., R. F. Thompson, *The Brain*, New York, W. H. Freeman and Co., 1985), and popular books. The articles in this section address two new areas of investigation into the visual system. The first, concerning brain mechanisms of visual attention is by Richard Wurtz, Michael Goldberg, and David Robinson of the National Eye Institute. They describe the primate visual system generally and then focus on the process by which the brain decides which visual objects in the world are significant.

The second article, by Tomaso Poggio of the Massachusetts Institute of Technology, concerns the new and rapidly growing field of computer vision. A fundamental problem in computer science is how to develop a computer system that can "see" the world as we do. The visual system of the brain provides a key model for computer scientists and computer programs, in turn, help us to understand the visual brain.

Animals "see" the world with both their eyes and ears. Indeed, some night hunters like the barn owl see the world better with their ears. In his article on the hearing of the barn owl, Eric Knudsen of Stanford University School of Medicine provides an overview of the barn owl's auditory system and how it develops.

7

Brain Mechanisms of Visual Attention

by Robert H. Wurtz, Michael E. Goldberg and David Lee Robinson
June 1982

The process by which the brain decides that certain objects in the world are significant is studied by recording the activity of nerve cells in the brain of monkeys that are responding to visual stimuli

Life is a bombardment of sights, sounds, smells, touches and tastes from which we select some information as a basis for action, ignoring all the rest. Thus we ignore the sensations evoked in our skin by our clothes but react immediately to the sensations evoked by an insect crawling under them. We ignore the din at a cocktail party but concentrate on our particular conversation. This ability to select objects of interest from the environment is called attention. William James described it well: "Everyone knows what attention is. It is the taking possession by the mind, in clear and vivid form, of one out of what seem several simultaneously possible objects or trains of thought. Focalization, concentration of consciousness are of its essence. It implies withdrawal from some things in order to deal effectively with others."

The process of visual attention, that is, the process of selecting important objects in the visual world, has been of particular interest to psychologists and physiologists because it has a reliable and measurable concomitant: the movement of the eyes. The reason is straightforward. The retina is distinctly nonuniform in its sensitivity to patterns of light. One tiny area of the retina, the fovea, which responds to light from the center of the visual field, has a high concentration of light-receptor cells; it therefore analyzes a pattern of light on a much finer scale than the periphery of the retina can. The brain rotates the eyes in their sockets so that the analytic power of the fovea is directed toward objects of interest. As you read these words, for example, your gaze jumps from one group of words to the next in a way that projects their image onto the fovea. Each such movement is called a saccade, from the French *saccader,* to jerk. Usually it lasts for less than 50 milliseconds. Nearly all the gathering of visual information that leads to visual perception occurs in the periods between saccades when the gaze is fixated on a visual target.

Since the process of visual attention is so closely linked to saccades, the question "What is a person attending to?" can be approximated by the question "What is the pattern of that person's eye movements?" Indeed, the Russian psychologist Alfred L. Yarbus, who recorded the patterns of people's eye movements as they looked at various pictures, has shown that the fixations tend to concentrate on a picture's most salient features. On the other hand, people also attend to objects they see "out of the corner of their eye." Michael T. Posner and his colleagues at the University of Oregon have done experiments in which the time it takes a person to respond to a target light by pressing a bar is taken to be an indication of how well the person attends to the light. People who know where the light will appear tend to respond more quickly to it even if they are expressly asked not to make a saccadic eye movement to it.

From these observations on visual attention one can devise a description of the process that makes it possible to investigate its mechanisms in the nervous system. First, visual attention implies the selection of a visual object from the visual field at the expense of other objects. Second, the selection must preserve the location of the object; obviously one must know where the attend-

ed object is. Third, the nature of the response to the selected object is not crucial; after all, one may gaze at an object, reach for it or simply notice it, but in each case one has attended to it.

In experiments that began at the National Institute of Mental Health and continued at the National Eye Institute we have searched for neural activity that meets these three criteria. Specifically, we have recorded the electrical activity of nerve cells in the brain of the rhesus monkey. We made these recordings at times when the monkeys were alert and able to move their eyes and attend to visual stimuli. We assume that for our purposes the monkey is an adequate model of man. One reason to think so is that the monkey's visual acuity, perception of color, perception of depth and pattern of saccades all resemble those of man. In addition the regions of the monkey's brain known to be involved with vision and with eye movement resemble their counterparts in the brain of man in both their position in the brain and their anatomical structure. We hope, therefore, that the results of anatomical, physiological and behavioral experiments with monkeys can yield some understanding of the neural basis of human behavior.

The attempt to record and interpret the electrical activity of nerve cells in the brain is facilitated by the fact that the activity has a characteristic pattern. Ordinarily a nerve cell has a resting electric potential: a voltage between the in-

EYE MOVEMENTS made by rhesus monkeys as they viewed pictures of the faces of other monkeys reveal patterns of visual attention. Here two monkeys' faces (*top row*) were each shown for eight seconds to monkeys named Joe and B.F. One face was more or less expressionless; the other face had an expression known as open-mouthed threat. The open mouth got much attention. In general Joe's visual fixations (*middle row*) and B.F.'s (*bottom row*) are connected by rapid eye movements, or saccades. The fixations trace out the faces' salient features. The illustration is based on the work of Caroline F. Keating of Colgate University and Gregory Keating of the Upstate Medical Center at Syracuse of the State University of New York.

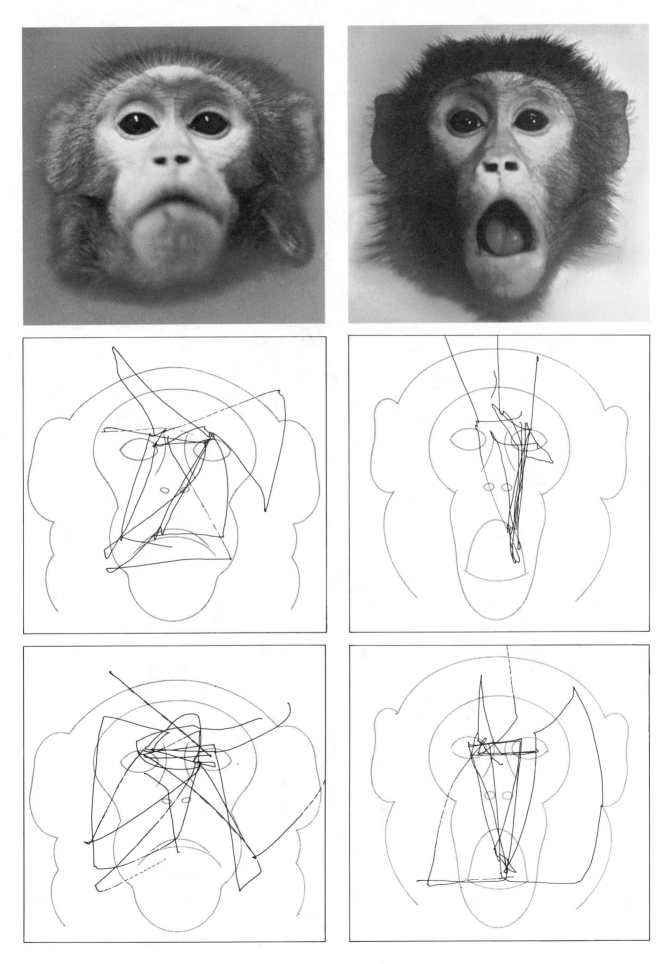

terior of the cell and the exterior. At certain times this potential changes dramatically; then a few milliseconds later it reverts to its resting value. The deviation from the resting potential, which is called an action potential, travels the length of the axon, the long fiber extending from the nerve cell. In this way the nerve cell sends messages to other nerve cells. The change in potential can readi-ly be measured by means of a fine-wire microelectrode placed near the cell.

Suppose a particular nerve cell is shown to generate a burst of action potentials whenever a spot of light impinges on a certain part of the retina. The cell discharges sporadically at other times, but even so it is fair to conclude that the cell is related to the neural mechanisms underlying visual perception or the ac-tions guided by vision. Similarly, suppose a nerve cell discharges in an intense and predictable way before a certain movement of the body. Presumably the cell is related to the neural generation of that movement.

In order to study the relation of nerve-cell activity to a monkey's behavior it is necessary to train the monkeys employed in our experiments in several visual tasks. First each monkey is taught to press a lever that makes a spot of light appear on a screen in front of it. A few seconds after the spot appears it dims and stays dim for half a second. If the monkey releases the lever while the spot is dim, it gets a drop of juice as a reward. The spot is so small and the dimming is so subtle that the monkey must train its fovea on it. (The people who try the task in our laboratory always do the same.) In fact, the monkey is sufficiently intent on watching the spot, which we call the fixation point, so that it seldom moves its eyes in saccades to look at spots of light we may flash elsewhere on the screen. If we move the fixation point, however, the monkey typically moves its eyes in a saccade from their former position to the new one.

When the monkey has learned these initial skills, it is placed under general anesthesia and surgically prepared for the making of neurophysiological recordings. We implant a device that restrains the monkey's head, sensors that measure its eye movements and a chamber through which microelectrodes can subsequently be introduced into the brain to monitor the activity of individual nerve cells. Then at the beginning of each day of experimentation the monkey is released from its home cage and allowed to climb into the experimental apparatus. Typically it does its tasks willingly. With a given monkey we usually study some hundreds of cells over a period of several months. Fortunately the brain is insensitive to pain. Indeed, the brain of human patients is often made the subject of electrical stimulation and recording while the patient is awake in order to render subsequent surgery (such as surgery to alleviate intractable epilepsy) more precise.

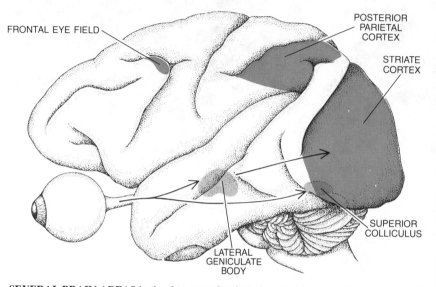

SEVERAL BRAIN AREAS in the rhesus monkey include cells that respond to visual stimuli (for example spots of light on a screen the monkey is watching) by changing their electrical activity. The superior colliculus gets signals direct from the retina. So does the lateral geniculate body, which communicates with the striate cortex. Two other cortical areas the authors studied also respond to visual stimuli. They are the frontal eye field and the posterior parietal cortex.

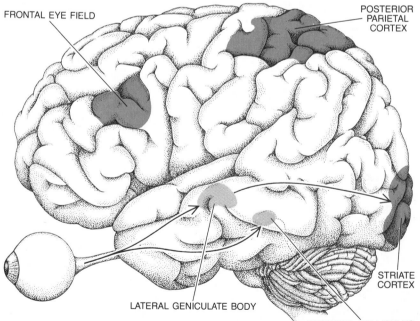

CORRESPONDING AREAS in the brain of man are diagrammed. Among them the superior colliculus, the lateral geniculate body and the striate cortex are markedly similar to their counterparts in the monkey. The great expanse of the cerebral cortex in the brain of man means, however, that the frontal eye field and the posterior parietal cortex are not easily delimited.

Since we wanted to study the physiology of visual attention, we began by examining the properties of nerve cells in areas of the brain that get signals from the retina by way of the optic nerve. In monkeys and in man the optic nerve divides into two major pathways. One of them leads to the lateral geniculate body, a collection of nerve cells whose axons project in turn to the striate cortex, the rearmost part of the cerebral cortex. Here the cortex begins the analysis of the visual world for color, shape, motion and depth.

The second main pathway leads to the

superior colliculus, an area in mammals at the top of the brain stem. The superior colliculus is the equivalent of the optic tectum, the area in less highly developed vertebrate animals that receives optic-nerve axons at the top of the brain stem. In animals such as frogs and fishes the optic tectum is the most important visual area.

We began, then, with cells in the superior colliculus. It was already well established that such cells, like the great majority of cells in the brain that respond to visual stimuli, respond not to diffuse light but to light falling on a specific area of the retina. We can map this area, which is called the receptive field of the cell, by measuring the response of the cell to spots of light flashed at various positions on the screen while the monkey is looking at the fixation point. Since the monkey keeps its gaze steady, we assume that the position of the spot of light on the screen with respect to the fixation point is a measure of the position of the image of the spot on the monkey's retina with respect to the fovea. Spots in some locations on the screen cause the cell to respond with a burst of action potentials; spots in other locations do not.

In the case of the superior colliculus it turns out that small spots of light flashed anywhere in a large part of the visual field (so that their image falls anywhere on a large part of the retina) can activate a cell. A receptive field near the fovea may correspond to an angle of four degrees in the visual field. Farther from the fovea the fields may each be 20 degrees in diameter. (The width of your forefinger at arm's length blocks out an angle of about one degree.)

While we were mapping the receptive fields of cells in the superior colliculus we made a fortuitous observation: whenever the monkey mistakenly glanced at a spot of light in the receptive field of the cell whose activity we were recording, the cell's response was much more intense than its response to such a spot if the monkey continued to fixate. We therefore modified our experiment so that the monkey was rewarded for these saccades with a drop of juice. We confirmed that many nerve cells in the superior colliculus discharge more intensely and more regularly when the spot of light in the receptive field is the target for an eye movement. In fact, roughly half of the cells in the superior colliculus that respond at all to visual stimuli showed this enhanced response. Typically the enhanced response begins about 50 milliseconds after the spot flashes onto the screen. The eye movement to the spot comes later: about 200 milliseconds after the spot appears.

Doubtless the monkey attended to the spot of light: its eyes moved in reaction to it. Doubtless too the vigorous dis-

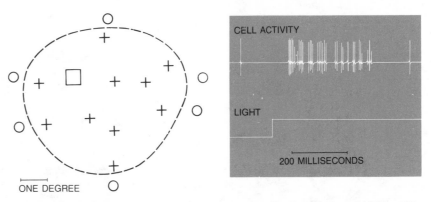

RECEPTIVE FIELD of a nerve cell is the specific part of the visual field to which the cell is sensitive. At the left a receptive field is mapped. Plus signs indicate points at which a spot of light made the cell discharge at an increased rate. Open dots indicate points at which a spot elicited no increase. The cell's response to a spot at the point marked by the square is at the right.

charge by cells in the monkey's superior colliculus implies that the activity of other brain cells (such as the ones to which the cells in the superior colliculus project their axons) was altered. Could the enhanced response therefore be part of a neural basis for a process by which "the mind takes possession of one object"?

In order to answer this question we had to solve several problems. First we had to determine whether the enhanced response required a visual stimulus or whether it was merely activity preceding a spontaneous eye movement. It did require a stimulus. When the eyes of the monkeys moved in saccades in total darkness, the cells did not discharge. The enhanced response was not simply a precursor of any eye movement; it was a modulation of the reaction to visual stimuli.

Next we had to determine whether the enhanced response could be associated with a subsequent saccade that had another stimulus as its target. It was conceivable that the effort of making any eye movement increases the excitability of all the cells in the brain that respond to visual stimuli. We examined this possibility by presenting the monkey simultaneously with two spots, one in the receptive field of the cell we were monitoring and one far outside it. When the monkey moved its eyes to the spot in the receptive field, the cell anticipated the saccade with an enhanced response. When the monkey moved its eyes to the spot outside the field, the response of the cell was not enhanced. This means the enhanced response was spatially selective: it is related to stimuli in a specific part of visual space.

Finally we had to test the possibility that the enhanced response came only when the monkey reacted to the stimulus in a particular way, that is, by making the stimulus the target of a saccade. In order to do so we collaborated with

Charles W. Mohler of our laboratory in designing an experiment in which monkeys reacted to spots of light in a different way. Again the monkey began by pressing a bar that made a peripheral spot of light appear. Here, however, the spot was large, and the monkey earned a reward by releasing the bar when either the spot or the fixation point dimmed. The fixation point was small enough and its degree of dimming was slight enough so that the monkey could perceive the dimming only with its fovea. In contrast, the dimming of the peripheral spot was evident to the monkey even if the monkey attended to the spot with the periphery of its retina. The monkey tended, therefore, to look at the fixation point, and when it reacted to the dimming of the peripheral spot, we could infer it had attended to the spot even though its eyes had not moved.

The experiment was conclusive: the cells in the superior colliculus that showed an enhanced response when the monkey made the spot the target of an eye movement gave no such response when the monkey attended to the spot by releasing the bar. It followed that the enhanced response in the superior colliculus was not a necessary concomitant of visual attention. More likely the enhancement is part of the neural processes involved in the initiation of eye movement.

In the other main visual pathway, the one that leads into the cerebral cortex, a similar sequence of experiments yielded quite different results. Typically a cell in the striate cortex showed only a slight increase in its rate of discharge when a spot of light in its receptive field was the target of a saccade. It showed this slight increase for a saccade to some other part of the visual field. Moreover, it showed the increase when the monkey attended to the spot of light by releasing the bar without an eye movement. Since the enhanced response in the striate cor-

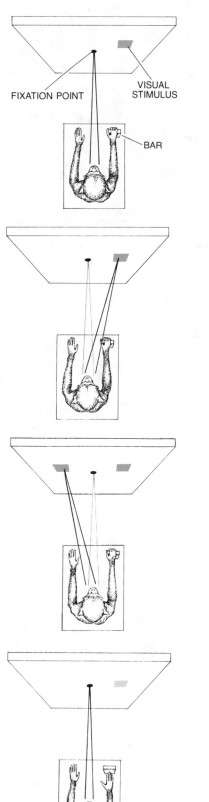

FIXATION POINT

VISUAL STIMULUS

BAR

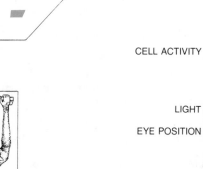

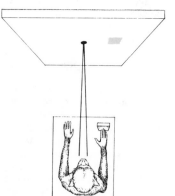

a

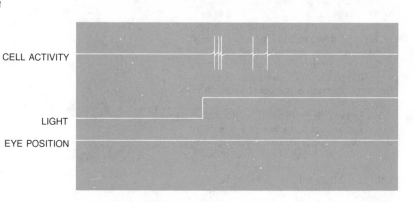

CELL ACTIVITY

LIGHT

EYE POSITION

b

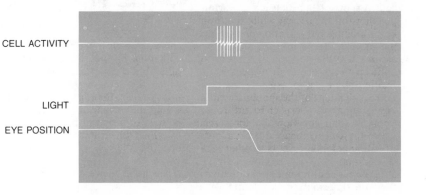

CELL ACTIVITY

LIGHT

EYE POSITION

c

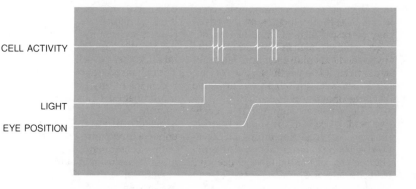

CELL ACTIVITY

LIGHT

EYE POSITION

d

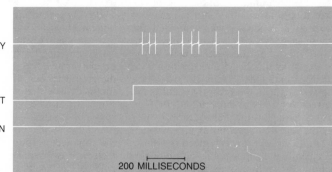

CELL ACTIVITY

LIGHT

EYE POSITION

200 MILLISECONDS

ACTIVITY OF SUPERIOR COLLICULUS when a spot of light appears on a screen depends on how the monkey reacts to the spot. For the experiments diagrammed here a cell in the superior colliculus was monitored as the spot was flashed in the cell's receptive field. In *a* the monkey kept its gaze directed at a central point of light (*fixation point*) and ignored the stimulus; the cell responded to the spot with a few bursts of activity (action potentials). In *b* the spot was the target for a saccade; the cell's activity was more pronounced. Two further experiments with the monkey failed to yield a similar enhanced activity. In *c* a spot outside the cell's receptive field was the target for the saccade. In *d* there was no saccade, but the monkey attended to a spot in the receptive field by releasing a bar when the spot dimmed.

tex is the same for any saccade, it does not show that the saccade is directed toward a stimulus in a given cell's receptive field. That is, the enhanced response is not spatially selective. Hence it cannot be related to visual attention. Perhaps it is related simply to how alert the monkey is. The monkey's state of alertness affects the brain's processing of all stimuli equally.

The results of the two sequences of experiments reinforce the differences already discovered between the two visual pathways. In the striate cortex the neural activity is strongly modulated by de-

tails of the pattern of light falling on the retina: the cells respond preferentially to slits of light at a particular place and orientation in the visual field. In addition the activation of some cells requires that the slit move in a certain direction or have a certain color. Our experiments show that these various responses are almost unchanged when the visual pattern is the target of a saccade or when it signals the monkey to release the bar. In other words, the striate cortex, the area of the cerebral cortex that is exquisitely sensitive to the details of the visual world, is essentially unaffected by the

significance the details may have for the monkey.

The superior colliculus is different. Here the cells that respond to visual stimuli perform a rather rudimentary computation: in general their activation requires only a small spot of light somewhere in a relatively large part of the visual field. In the superior colliculus, however, the response of the cells is modulated quite clearly when the monkey makes the spot of light the target of a saccade. The visual processing may be rudimentary, but the cells are sensitive to the significance of the stimulus.

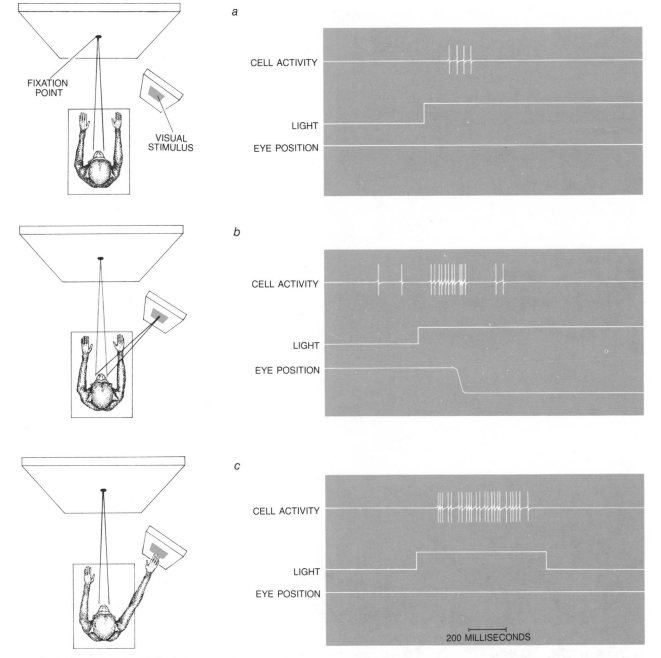

ACTIVITY OF POSTERIOR PARIETAL CORTEX differs from the activity of the superior colliculus (and also the striate cortex and the frontal eye field) in that enhanced activity was found when the monkey attended to the spot of light regardless of how it attended to it. In *a* the monkey showed no reaction; a posterior parietal cell responded to the spot with a few action potentials. In *b* the spot was the target for a saccade; the cell's activity was enhanced. In *c* the monkey touched the spot, and again the cell's activity was enhanced.

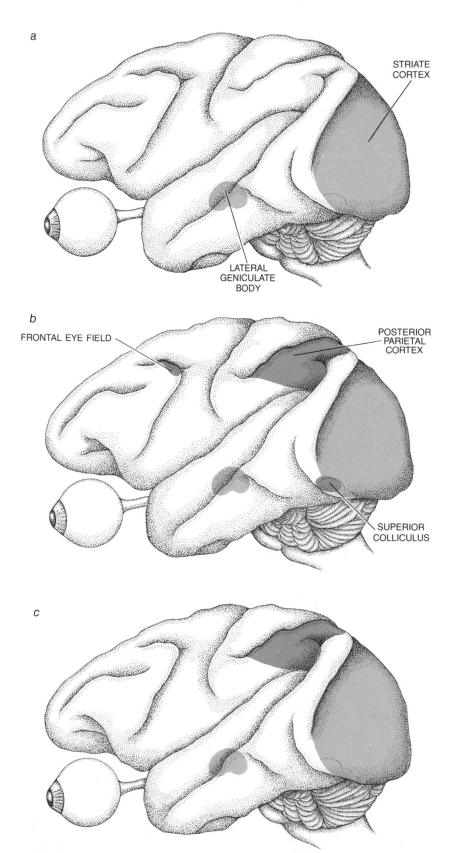

a

STRIATE
CORTEX

LATERAL
GENICULATE
BODY

b

FRONTAL EYE FIELD

POSTERIOR
PARIETAL
CORTEX

SUPERIOR
COLLICULUS

c

PATTERNS OF ENHANCED ACTIVITY in the areas of the brain responsive to visual stimuli can be surmised from the authors' experiments at the National Institutes of Health. Visual stimuli activate the pathways leading through the lateral geniculate body to the striate cortex (*a*). Visual stimuli that become the target for a saccade activate preferentially the superior colliculus, the frontal eye field and the posterior parietal cortex (*b*). Visual stimuli to which the monkey attends in any manner activate preferentially the posterior parietal cortex (*c*). This suggests that the posterior parietal cortex is part of the neural substrate for visual attention.

Our next experiments showed that cells in an area of the cerebral cortex respond to visual stimuli much as the cells in the superior colliculus do. The area lies in the frontal lobe in front of the motor cortex, a region important for the control of movements of the body. In the 19th century the Scottish physiologist Sir David Ferrier showed that in monkeys the electrical stimulation of this area of the cortex evoked eye movements. He called the area the frontal eye field. For some time investigators assumed that the frontal eye field controlled eye movements much as the nearby motor cortex controlled the movements of the rest of the body. Then, however, Emilio Bizzi, who was working at the National Institute of Mental Health, made electrical recordings of the activity of nerve cells in the frontal eye field. He could find none that reliably discharged in advance of a monkey's spontaneous eye movements. A few of the cells discharged during and after eye movements, but these cells could not, of course, be initiators of eye movement.

When we ourselves studied the frontal eye field, we found that many of the cells in the area discharge in response to small spots of light appearing anywhere in a receptive field that can be as large as a fourth of the entire visual field. (A receptive field of that size is larger than the receptive field of a typical cell in the superior colliculus and many times larger than the field of a typical cell in the striate cortex.) The nerve cells in the frontal eye field of the monkey showed no response anticipating spontaneous saccades in the dark. On the other hand, they showed an enhanced response when a spot of light in a given cell's receptive field was the target for a saccade. They did not show an enhanced response before a saccade that directed the gaze somewhere else or before the monkey attended to the spot without making a saccade.

It follows that these nerve cells, like the nerve cells in the superior colliculus, cannot account for visual attention. The reason is the same: attention can be dissociated from eye movement but these cells' enhanced discharge cannot. Perhaps, however, the nerve cells we have studied in the frontal eye field account for Ferrier's finding. When we apply a small electric current to a cell in the frontal eye field that is responsive to visual stimuli, a saccade moves the eyes to the center of the cell's receptive field. We assume the cell normally supplies visual information to the neural mechanisms that make the eyes move; the electrical stimulation passes along this channel of communication so that the eyes of the monkey move in a saccade as if there has been a visual target.

If the superior colliculus, the striate cortex and the frontal eye field do

SELF-PORTRAITS made by the German artist Anton Räderscheidt are evidence that the posterior parietal cortex is involved in brain mechanisms of visual attention. The first self-portrait (*upper left*) was made two months after Räderscheidt suffered a stroke that damaged the parietal cortex on the right side of his brain. Half of the face is omitted and half of the paper is blank. It is as if the artist could not attend to the left half of the world. The second self-portrait (*upper right*) was made three and a half months after the stroke, the third one (*lower left*) six months after the stroke and the fourth one (*lower right*) nine months after the stroke. The self-portraits were collected by Richard Jung of the University of Freiburg and are published through the courtesy of Gisele Räderscheidt, the widow of the artist.

not mediate visual attention, what does? Clinical evidence implicates a part of the cerebral cortex designated the posterior parietal cortex. Specifically, people with damage to the posterior parietal cortex on the right side of the brain tend to ignore objects in the left half of their visual field. Such people can see the objects, but they do not attend to them. A striking example of this deficiency is seen in a series of self-portraits made by the German artist Anton Räderscheidt after he had suffered a stroke that damaged the parietal cortex on the right side of his brain [see illustration on page 89]. A self-portrait he made early in his recovery from the stroke omits the left side of his face.

Experiments done with monkeys by Juhani Hyvärinen and his colleagues at the University of Helsinki have shown that cells in the posterior parietal cortex are active when the monkey's behavior is guided by vision, for example when the monkey reaches for an object or tracks the motion of an object with its eyes. Experiments done by Vernon B. Mountcastle and his colleagues at the Johns Hopkins University School of Medicine have shown that cells in the posterior parietal cortex discharge just before eye movements, and Mountcastle has proposed that the activity of such cells is related to visual attention. On the basis of these efforts and the clinical evidence we decided (with Gregory Stanton and Catherine Bushnell of our laboratory) to make the posterior parietal cortex a subject of our own experiments. We found that cells there have receptive fields similar to the receptive fields of the cells in the frontal eye fields. The cells in the posterior parietal cortex respond best, however, to large stimuli, for example spots of light five degrees in diameter.

When our monkeys made such a stimulus the target for a saccade, the cells we were monitoring in the posterior parietal cortex showed an enhanced response. The response was spatially selective: we observed it when the saccade was directed to a stimulus in the cell's receptive field but not when the saccade went elsewhere. In this respect the enhanced response resembled the one we have noted in the superior colliculus and in the frontal eye field.

We next required the monkey to gaze at a fixation point and attend to the dimming of a peripheral stimulus by releasing a bar. We found an enhanced response. To confirm that this enhancement in the posterior parietal cortex is independent of eye movement we devised another task. As before, the monkey was required to gaze at the fixation point. On some trials, however, the fixation point dimmed; the monkey released the bar to earn its reward. On other trials the monkey had to touch an illuminated panel without taking its eyes off the fixation point. If the panel was in the receptive field of the cell we were monitoring, the cell showed an enhanced response. The cell's activity was as intense as it had been when the monkey made a saccade to the panel.

The enhanced response of cells in the posterior parietal cortex is spatially selective, and it is independent of the particular action the monkey takes toward the stimulus. This makes it different from the enhanced response of cells in the superior colliculus and the frontal eye field. It also fits the clinical evidence that the posterior parietal cortex is involved in visual attention.

The phenomenon of enhancement suggests the outlines of a neural mechanism by which the monkey chooses objects from the visual world and makes them the targets for saccades. The retina can be taken to send code to stations in the brain describing the visual world. If the monkey is alert but not attending to anything in particular, the response of the nerve cells in these stations is relatively uniform. When the monkey begins to attend to some object, however, the nerve cells in the posterior parietal cortex that are related to the object because it is in their receptive field begin to discharge more intensely. If the monkey decides to make a saccade toward the object in order to examine it more closely, the cells in the superior colliculus and in the frontal eye field that are related to the object will also discharge intensely. The enhanced response has changed the uniform state of activity in the visual areas into a state in which the activity of nerve cells responding to one object is greater than the activity of nerve cells responding to other objects. This change happens even if the selected object is no more striking than the rest of the visual field. The only difference is that the monkey has decided to attend to the object.

We do not know how the enhancement arises, and in the absence of this knowledge it is tempting to say that the enhanced response amounts in itself to visual attention. It remains quite possible, however, that the enhancement is only a correlate of visual attention. That is, it may accompany visual attention, just as an eye movement may, but not be part of the neural mechanism whose product is attention. Nevertheless, the demonstration of brain activity whose enhancement is related to attention means we have at least begun to translate a psychological concept into physiological terms.

8

Vision by Man and Machine

by Tomaso Poggio
April 1984

*How does an animal see? How might a computer do it?
A study of stereo vision guides research on both these
questions. Brain science suggests computer programs;
the computer suggests what to look for in the brain*

The development of computers of increasing power and sophistication often stimulates comparisons between them and the human brain, and these comparisons are becoming more earnest as computers are applied more and more to tasks formerly associated with essentially human activities and capabilities. Indeed, it is widely expected that a coming generation of computers and robots will have sensory, motor and even "intellectual" skills closely resembling our own. How might such machines be designed? Can our rapidly growing knowledge of the human brain be a guide? And at the same time can our advances in "artificial intelligence" help us to understand the brain?

At the level of their hardware (the brain's or a computer's) the differences are great. The neurons, or nerve cells, in a brain are small, delicate structures bound by a complex membrane and closely packed in a medium of supporting cells that control a complex and probably quite variable chemical environment. They are very unlike the wires and etched crystals of semiconducting materials on which computers are based. In the organization of the hardware the differences also are great. The connections between neurons are very numerous (any one neuron may receive many thousands of inputs) and are distributed in three dimensions. In a computer the wires linking circuit components are limited by present-day solid-state technology to a relatively small number arranged more or less two-dimensionally.

In the transmission of signals the differences again are great. The binary (on-off) electric pulses of the computer are mirrored to some extent in the all-or-nothing signal conducted along nerve fibers, but in addition the brain employs graded electrical signals, chemical messenger substances and the transport of

ions. In temporal organization the differences are immense. Computers process information serially (one step at a time) but at a very fast rate. The time course of their operation is governed by a computer-wide clock. What is known of the brain suggests that it functions much slower but that it analyzes information along millions of channels concurrently without need of clock-driven operation.

How, then, are brains and computers alike? Clearly there must be a level at which any two mechanisms can be compared. One can compare the tasks they do. "To bring the good news from Ghent to Aix" is a description of a task that can be done by satellite, telegraph, horseback messenger or pigeon post equally well (unless other constraints such as time are specified). If, therefore, we assert that brains and computers function as information-processing systems, we can develop descriptions of the tasks they perform that will be equally applicable to either. We shall have a common language in which to discuss them: the language of information processing. Note that in this language descriptions of tasks are decoupled from descriptions of the hardware that perform them. This separability is at the foundation of the science of artificial intelligence. Its goals are to make computers more useful by endowing them with "intelligent" capabilities, and beyond that to understand the principles that make intelligence possible.

In no field have the descriptions of information-processing tasks been more precisely formulated than in the study of vision. On the one hand it is the dominant sensory modality of human beings. If we want to create robots capable of performing complex manipulative tasks in a changing environment, we must surely endow them with adequate visual powers. Yet vision remains elusive. It is something we are good at; the

brain does it rapidly and easily. It is nonetheless a mammoth information-processing task. If it required a conscious effort, like adding numbers in our head, we would not undervalue its difficulty. Instead we are easily lured into oversimple, noncomputational preconceptions of what vision really entails.

Ultimately, of course, one wants to know how vision is performed by the biological hardware of neurons and their synaptic interconnections. But vision is not exclusively a problem in anatomy and physiology: how nerve cells are interconnected and how they act. From the perspective of information processing (by the brain or by a computer) it is a problem at many levels: the level of computation (What computational tasks must a visual system perform?), the level of algorithm (What sequence of steps completes the task?) and then the level of hardware (How might neurons or electronic circuits execute the algorithm?). Thus an attack on the problem of vision requires a variety of aids, including psychophysical evidence (that is, knowledge of how well people can see) and neurophysiological data (knowledge of what neurons can do). Finding workable algorithms is the most critical part of the project, because algorithms are constrained both by the computation and by the available hardware.

Here I shall outline an effort in which I am involved, one that explores a sequence of algorithms first to extract information, notably edges, or pronounced contours, in the intensity of light, from visual images and then to calculate from those edges the depths of objects in the three-dimensional world. I shall concentrate on a particular aspect of the task, namely stereopsis, or stereo vision. Not the least of my reasons is the central role stereopsis has played in the work on vision that my colleagues and I have done at the Artificial Intelligence

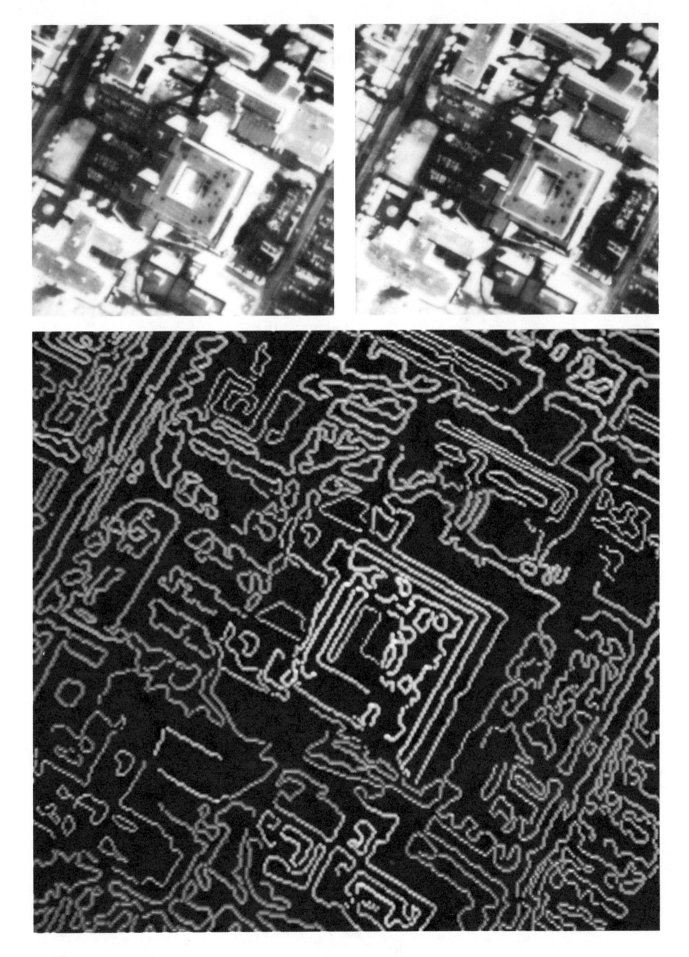

Laboratory of the Massachusetts Institute of Technology. In particular, stereopsis has stimulated a close investigation of the very first steps in visual information processing. Then too, stereopsis is deceptively simple. As with so many other tasks that the brain performs without effort, the development of an automatic system with stereo vision has proved to be surprisingly difficult. Finally, the study of stereopsis benefits from the availability of a large body of psychophysical evidence that defines and constrains the problem.

The information available at the outset of the process of vision is a two-dimensional array of measurements of the amount of light reflected into the eye or into a camera from points on the surfaces of objects in the three-dimensional visual world. In the human eye the measurements are made by photoreceptors (rod cells and cone cells), of which there are more than 100 million. In a camera that my colleagues and I use at the Artificial Intelligence Laboratory the processes are different but the result is much the same. There the measurements are made by solid-state electronic sensors. They produce an array of 1,000 by 1,000 light-intensity values. Each value is a pixel, or picture element.

In either case it is inconceivable that the gap between the raw image (the large array of numbers produced by the eye or the camera) and vision (knowing *what* is around, and *where*) can be spanned in a single step. One concludes that vision requires various processes—one thinks of them as modules—operating in parallel on raw images and producing intermediate representations of the images on which other processes can work. For example, several vision modules seem to be involved in reconstructing the three-dimensional geometry of the world. A short list of such modules would have to include modules that deduce shape from shading, from visual texture, from motion, from contours, from occlusions and from stereopsis. Some may work directly on the raw image (the intensity measurements). Of-

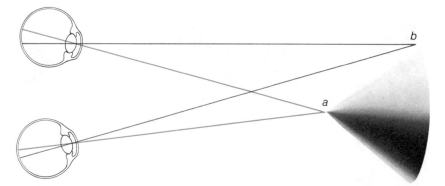

BINOCULAR DISPARITIES are the basis for stereopsis. They arise because the eyes converge slightly, so that their axes of vision meet at a point in the external world (*a*). The point is "fixated." A neighboring point in the world (*b*) will then project to a point on the retina some distance from the center of vision. The distance will not be the same for each eye.

ten, however, a module may operate more effectively on an intermediate representation.

Stereopsis arises from the fact that our two eyes view the visual world from slightly different angles. To put it another way, the eyes converge slightly, so that their axes of vision meet at a point in the visual world. The point is said to be fixated by the eyes, that is, the image of the point falls on the center of vision of each retina. Any neighboring point in the visual field will then project to a point on each retina some distance from the center of vision. In general this distance will not be the same for both eyes. In fact, the disparity in distance will vary with the depth of the point in the visual field with respect to the fixated point.

Stereopsis, then, is the decoding of three-dimensionality from binocular disparities. It might appear at first to be a straightforward problem in trigonometry. One might therefore be tempted to program a computer to solve it that way. The effort would fail; our own facility with stereopsis has led us to gloss over the central difficulty of the task, as we may see if we formally set out the steps involved in the task. They are four: A location in space must be selected from

one retinal image. The same location must be identified in the other retinal image. Their positions must be measured. From the disparity between the two measurements the distance to the location must be calculated.

The last two steps are indeed an exercise in trigonometry (at least in the cases considered in this article). The first two steps are different. They require, in effect, that the projection of the same point in the physical world be found in each eye. A group of contiguous photoreceptors in one eye can be thought of as looking along a line of sight to a patch of the surface of some object. The photoreceptors looking at the same patch of surface from the opposite eye must then be identified. Because of binocular disparity they will not be at the same position with respect to the center of vision.

This, of course, is where the difficulty lies. For us the visual world contains surfaces that seem effectively labeled because they belong to distinct shapes in specific spatial relations to one another. One must remember, however, that vision begins with no more than arrays of raw light intensity measured from point to point. Could it be that the brain matches patterns of raw light intensity from one eye to the other? Probably not. Experiments with computers place limits on the effectiveness of the matching, and physiological and psychophysical evidence speaks against it for the human visual system. For one thing, a given patch of surface will not necessarily reflect the same intensity of light to both eyes. More important, patches of surface widely separated in the visual world may happen to have the same intensity. Matching such patches would be incorrect.

A discovery made at AT&T Bell Laboratories by Bela Julesz shows the full extent of the problem. Julesz devised pairs of what he called random-dot stereograms. They are visual stimuli that

STEREO VISION BY A COMPUTER exemplifies the study of vision as a problem in information processing. The images at the top of the opposite page are aerial photographs provided by Robert J. Woodham of the University of British Columbia. They show part of the university's campus. In two ways they mimic the visual data on which biological vision is based. First, they were made from different angles, so that objects in one image have a slightly different position in the other. The two eyes of human beings also see the world from different angles. Second, the images were made by a mosaic of microelectronic sensors, each of which measures the intensity of light along a particular line of sight. The photoreceptor cells of the eye do much the same thing. The map at the bottom was generated by a computer programmed to follow an algorithm, or procedure, devised by David Marr and the author at the Artificial Intelligence Laboratory of the Massachusetts Institute of Technology and further developed there by W. Eric L. Grimson. The computer filtered the images to emphasize spatial changes in intensity. Then it performed stereopsis: it matched features from one image to the other, determined the disparity between their positions and calculated their relative depths in the three-dimensional world. Increasing elevations in the map are coded in colors from blue to red.

225	221	216	219	219	214	207	218	219	220	207	155	136	135	130	131	125
213	206	213	223	208	217	223	221	223	216	195	156	141	130	128	138	123
206	217	210	216	224	223	228	230	234	216	207	157	136	132	137	130	128
211	213	221	223	220	222	237	216	219	220	176	149	137	132	125	136	121
216	210	231	227	224	228	231	210	195	227	181	141	131	133	131	124	122
223	229	218	230	228	214	213	209	198	224	161	140	133	127	133	122	133
220	219	224	220	219	215	215	206	206	221	159	143	133	131	129	127	127
221	215	211	214	220	218	221	212	218	204	148	141	131	130	128	129	118
214	211	211	218	214	220	226	216	223	209	143	141	141	124	121	132	125
211	208	223	213	216	226	231	230	241	199	153	141	136	125	131	125	136
200	224	219	215	217	224	232	241	240	211	150	139	128	132	129	124	132
204	206	208	205	233	241	241	252	242	192	151	141	133	130	127	129	129
200	205	201	216	232	248	255	246	231	210	149	141	132	126	134	128	139
191	194	209	238	245	255	249	235	238	197	146	139	130	132	129	132	123
189	199	200	227	239	237	235	236	247	192	145	142	124	133	125	138	128
198	196	209	211	210	215	236	240	232	177	142	137	135	124	129	132	128
198	203	205	208	211	224	226	240	210	160	139	132	129	130	122	124	131
216	209	214	220	210	231	245	219	169	143	148	129	128	136	124	128	123
211	210	217	218	214	227	244	221	162	140	139	129	133	131	122	126	128
215	210	216	216	209	220	248	200	156	139	131	129	139	128	123	130	128
219	220	211	208	205	209	240	217	154	141	127	130	124	142	134	128	129
229	224	212	214	220	229	234	208	151	145	128	128	142	122	126	132	124
252	224	222	224	233	244	228	213	143	141	135	128	131	129	128	124	131
255	235	230	249	253	240	228	193	147	139	132	128	136	125	125	128	119
250	245	238	245	246	235	235	190	139	136	134	135	126	130	126	137	132
240	238	233	232	235	255	246	168	156	141	129	127	136	134	135	130	126
241	242	225	219	225	255	255	183	139	141	126	139	128	137	128	128	130
234	218	221	217	211	252	242	166	144	139	132	130	128	129	127	121	132
231	221	219	214	218	225	238	171	145	141	124	134	131	134	131	126	131
228	212	214	214	213	208	209	159	134	136	139	134	126	127	127	124	122
219	213	215	215	205	215	222	161	135	141	128	129	131	128	125	128	127

BEGINNING OF VISION for an animal or a computer is a gray-level array: a point-by-point representation of the intensity of light produced by a grid of detectors in the eye or in a digital camera. The image at the top of this illustration is such an array. It was produced by a digital camera as a set of intensity values in a grid of 576 by 454 picture elements ("pixels"). Intensity values for the part of the image inside the rectangle are given digitally at the bottom.

contain no perceptual clues except binocular disparities. To make each pair he generated a random texture of black and white dots and made two copies of it. In one of the copies he shifted an area of the pattern, say a square. In the other copy he shifted the square in the opposite direction. He filled the resulting hole in each pattern with more random texture. Viewed one at a time each such pattern looked uniformly random. Viewed through a stereoscope, so that each eye saw one of the patterns and the brain could fuse the two, the result was startling. The square gave a vivid impression of floating in front of its surroundings or behind them. Evidently stereopsis does not require the prior perception of objects or the recognition of shapes.

Julesz' discovery enables one to formulate the computational goal of stereopsis: it is the extraction of binocular disparities from a pair of images without the need for obvious monocular clues. In addition the discovery enables one to formulate the computational problem inherent in stereopsis. It is the correspondence problem: the matching of elements in the two images that correspond to the same location in space without the recognition of objects or their parts. In random-dot stereograms the black dots in each image are all the same: they have the same size, the same shape and the same brightness. Any one of them could in principle be matched with any one of a great number of dots in the other image. And yet the brain solves this false-target dilemma: it consistently chooses only the correct set of matches.

It must use more than the dots themselves. In particular, the fact that the brain can solve the correspondence problem shows it exploits a set of implicit assumptions about the visual world, assumptions that constrain the correspondence problem, making it determined and solvable. In 1976 David Marr and I, working at M.I.T., found that simple properties of physical surfaces could limit the problem sufficiently for the stereopsis algorithms (procedures to be followed by a computer) we were then investigating. These are, first, that a given point on a physical surface has only one three-dimensional location at any given time and, second, that physical objects are cohesive and usually are opaque, so that the variation in depth over a surface is generally smooth, with discontinuous changes occurring only at boundary lines. The first of these constraints—uniqueness of location—means that each item in either image (say each dot in a random-dot stereogram) has a unique disparity and can be matched with no more than one item in the other image. The second constraint—continuity and opacity—means that disparity varies smoothly except at object boundaries.

Together the two constraints provide matching rules that are reasonable and powerful. I shall describe some simple ones below. Before that, however, it is necessary to specify the items to be matched. After all, the visual world is not a random-dot stereogram, consisting only of black and white dots. We have already seen that intensity values are too unreliable. Yet the information the brain requires is encrypted in the intensity array provided by photoreceptors. If an additional property of physical surfaces is invoked, the problem is simplified. It is based on the observation that at places where there are physical changes in a surface, the image of the surface usually shows sharp variations in intensity. These variations (caused by markings on a surface and by variations in its depth) would be more reliable tokens for matching than raw intensities would be.

Instead of raw numerical values of intensity, therefore, one seeks a more symbolic, compact and robust representation of the visual world: a description of the world in which the primitive symbols—the signs in which the visual world is coded—are intensity variations. Marr called it a "primal sketch." In essence it is the conversion of the gray-level arrays provided by the visual photoreceptors into a form that makes explicit the position, direction, scale and magnitude of significant light-intensity gradients, with which the brain's stereopsis module can solve the correspondence problem and reconstruct the three-dimensional geometry of the visual world. I shall describe a scheme we have been using at the Artificial Intelligence Laboratory for the past few years, based on old and new ideas developed by a number of investigators, primarily Marr, Ellen C. Hildreth and me. It has several attractive features: it is fairly simple, it works well and it shows interesting resemblances to biological vision, which, in fact, suggested it. It is not, however, the full solution. Perhaps it is best seen as a working hypothesis about vision.

Basically the changes of intensity in an image can be detected by comparing neighboring intensity values in the image: if the difference between them is great, the intensity is changing rapidly. In mathematical terms the operation amounts to taking the first derivative. (The first derivative is simply the rate of change of a mathematical function. Here it is simply the rate at which intensity changes on a path across the gray-level array.) The position of an extremal value—a peak or a pit—in the first derivative turns out to localize the position of an intensity edge quite well [see illustra-

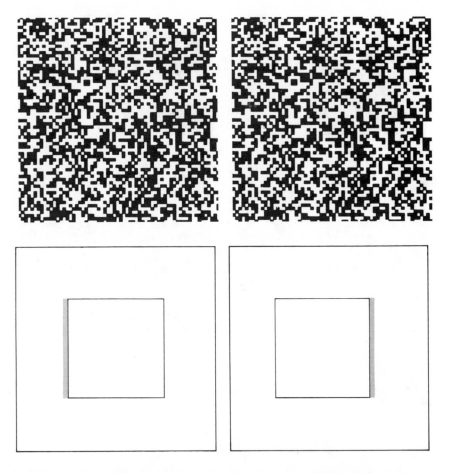

RANDOM-DOT STEREOGRAMS devised by Bela Julesz of AT&T Bell Laboratories are visual textures containing no clues for stereo vision except binocular disparities. The stereograms themselves are the same random texture of black and white dots (top). In one of them, however, a square of the texture is shifted toward the left; in the other it is shifted toward the right (bottom). The resulting hole in each image is filled with more random dots (gray areas).

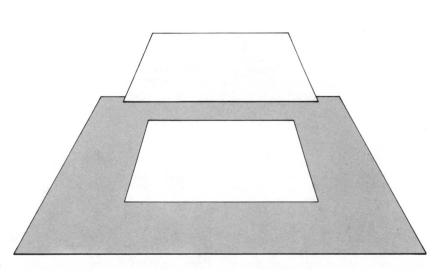

VIVID PERCEPTION OF DEPTH results when the random-dot stereograms shown in the bottom illustration on the preceding page are viewed through a stereoscope, so that each eye sees one of the pair and the brain can fuse the two. The sight of part of the image "floating" establishes that stereopsis does not require the recognition of objects in the visual world.

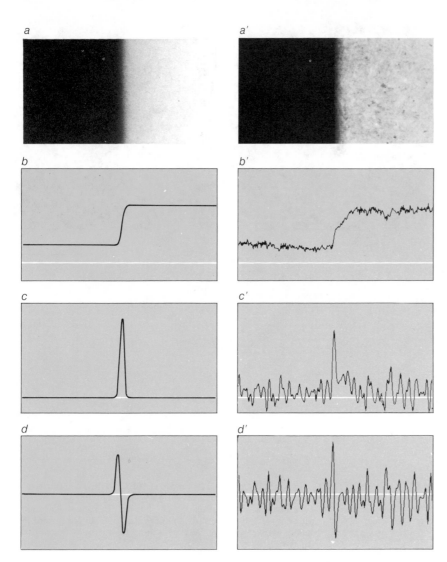

a

b

c

d

a'

b'

c'

d'

SPATIAL DERIVATIVES of an image serve to emphasize its spatial variations in intensity. The left part of the illustration shows an edge between two even shades of gray (*a*). The intensity along a path across the edge appears below it (*b*). The first derivative of the intensity is the rate at which intensity changes (*c*). Toward the left or toward the right there is no change; the first derivative therefore is zero. Along the edge itself, however, the rate of change rises and falls. The second derivative of the intensity is the rate of change of the rate of change (*d*). Both derivatives emphasize the edge. The first derivative marks it with a peak; the second derivative marks it by crossing zero. The right part of the illustration shows an edge more typical of the visual world (*a'*). The related intensity contour (*b'*) and its first and second derivatives (*c'*, *d'*) are "noisy." The edge must be smoothed before derivatives are taken. This illustration and the one on page 94 were prepared by H. Keith Nishihara of the Artificial Intelligence Laboratory.

tion above on this page]. In turn the intensity edge often corresponds to an edge on a physical surface. The second derivative also serves well. It is simply the rate of change of the rate of change and is obtained by taking differences between neighboring values of the first derivative. In the second derivative an intensity edge in the gray-level array corresponds to a zero-crossing: a place where the second derivative crosses zero as it falls from positive values to negative values or rises from negative values to positive.

Derivatives seem quite promising. Used alone, however, they seldom work on a real image, largely because the intensity changes in a real image are rarely clean and sharp changes from one intensity value to another. For one thing, many different changes, slow and fast, often overlap on a variety of different spatial scales. In addition changes in intensity are often corrupted by the visual analogue of noise. They are corrupted, in other words, by random disruptions that infiltrate at different stages as the image formed by the optics of the eye or

of a camera is transduced into an array of intensity measurements. In order to cope both with noisy edges and with edges at different spatial scales the image must be "smoothed" by a local averaging of neighboring intensity values. The differencing operation that amounts to the taking of first and second derivatives can then be performed.

There are various ways the sequence can be managed, and much theoretical effort has gone into the search for optimal methods. In one of the simplest the two operations—smoothing and differentiation—are combined into one. In technical terms it sounds forbidding: the image is convolved with a filter that embodies a particular center-surround function, the Laplacian of a Gaussian. It is not as bad as it sounds. A two-dimensional Gaussian is the bell-shaped distribution familiar to statisticians. In this context it specifies the importance to be assigned to the neighborhood of each pixel when the image is being smoothed. As the distance increases, the importance decreases. A Laplacian is a second derivative that gives equal weight to all paths extending away from a point. The Laplacian of a Gaussian converts the bell-shaped distribution into something more like a Mexican hat. The bell is narrowed and at its sides a circular negative dip develops.

Now the procedure can be described nontechnically. Convolving an image with a filter that embodies the Laplacian of a Gaussian is equivalent to substituting for each pixel in the image a weighted average of neighboring pixels, where the weights are provided by the Laplacian of a Gaussian. Thus the filter is applied to each pixel. It assigns the greatest positive weight to that pixel and decreasing positive weights to the pixels nearby [*see illustration on pages 98 and 99*]. Then comes an annulus—a ring—in which the pixels are given negative weightings. Bright points there feed negative numbers into the averaging. The result of the overall filtering is an array of positive and negative numbers: a kind of second derivative of the image intensity at the scale of the filter. The zero-crossings in this filtered array correspond to places in the original image where its intensity changed most rapidly. Note that a binary (that is, a two-valued) map showing merely the positive and negative regions of the filtered array is essentially equivalent to a map of the zero-crossings in that one can be constructed from the other.

In the human brain most of the hardware required to perform such a filtering seems to be present. As early as 1865 Ernst Mach observed that visual perception seems to enhance spatial variations in light intensity. He postulated that the enhancement might be achieved by lateral inhibition, a brain mechanism in which the excitation of an axon, or nerve fiber, say by a spot of bright light in the visual world, blocks the excitation of neighboring axons. The operation plainly enhances the contrast between the bright spot and its surroundings. Hence it is similar to the taking of a spatial derivative.

Then in the 1950's and 1960's evidence accumulated suggesting that the retina does something much like center-surround filtering. The output from each retina is conveyed to the rest of the brain by about a million nerve fibers, each being the axon of a neuron called a retinal ganglion cell. The cell derives its input (by way of intermediate neurons) from a group of photoreceptors, which form a "receptive field." What the evidence suggests is that for certain ganglion cells the receptive field has a center-surround organization closely approximating the Laplacian of a Gaussian. Brightness in the center of the receptive field excites the ganglion cell; brightness in a surrounding annulus inhibits it. In short, the receptive field has an ON-center and an OFF-surround, just like the Mexican hat.

Other ganglion cells have the opposite properties: they are OFF-center, ON-surround. If axons could signal negative numbers, these cells would be redundant: they report simply the negation of what the ON-center cells report. Neurons, however, cannot readily transmit negative activity; the ones that transmit all-or-nothing activity are either active or quiescent. Nature, then, may need neuronal opposites. Positive values in an image subjected to center-surround filtering could be represented by the activity of ON-center cells; negative values could be represented by the activity of OFF-center cells. In this regard I cannot refrain from mentioning the recent finding that ON-center and OFF-center ganglion cells are segregated into two different layers, at least in the retina of the cat. The maps generated by our computer might thus depict neural activity rather literally. In the maps on the opposite page red might correspond to ON-layer activity and blue to OFF-layer activity. Zero-crossings (that is, transitions from one color to the other) would be the locations where activity switches from one layer to the other. Here, then, is a conjecture linking a computational theory of vision to the brain hardware serving biological vision.

It should be said that the center-surround filtering of an image is computationally expensive for a computer because it involves great numbers of multiplications: about a billion for an image of 1,000 pixels by 1,000. At the Artificial Intelligence Laboratory, H. Keith Nishihara and Noble G. Larson, Jr., have designed a specialized device: a convolver

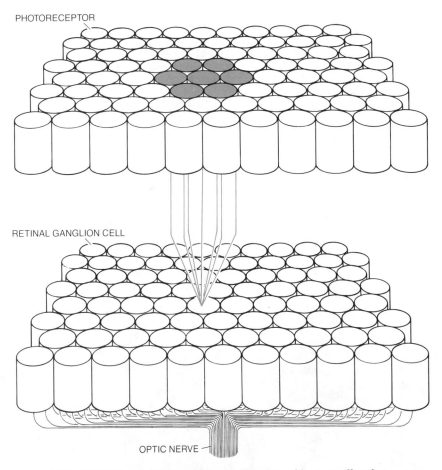

PHOTORECEPTOR

RETINAL GANGLION CELL

OPTIC NERVE

BIOLOGICAL FILTER embodied by cells in the retina resembles in its effect the computer procedure shown in the illustration on the following two pages. The filter begins with a layer of photoreceptors, which measure the light intensity of the visual world. They are connected by way of intermediate nerve cells not shown in the diagram to a layer of retinal ganglion cells, which send visual data to higher visual centers of the brain. For the sake of simplicity only one set of connections is shown. A photoreceptor cell (*red*) excites an "ON center" ganglion cell by promoting its tendency to generate neural signals; the surrounding photoreceptors (*blue*) inhibit the ganglion cell. The arrangement amounts to biological center-surround filtering.

that performs the operation in about a second. The speed is impressive but is plodding compared with that of the retinal ganglion cells.

I should also mention the issue of spatial scale. In an image there are fine changes in intensity as well as coarse. All must be detected and represented. How can it be done? The natural solution (and the solution suggested by physiology and psychophysics) is to use center-surround filters of different sizes. The filters turn out to be band-pass: they respond optimally to a certain range of spatial frequencies. In other words, they "see" only changes in intensity from pixel to pixel that are neither too fast nor too slow. For any one spatial scale the process of finding intensity changes consists, therefore, of filtering the image with a center-surround filter (or receptive field) of a particular size and then finding the zero-crossings in the filtered

image. For a combination of scales it is necessary to set up filters of different sizes, performing the same computation for each scale. Large filters would then detect soft or blurred edges as well as overall illumination changes; small filters would detect finer details. Sharp edges would be detected at all scales.

Recent theoretical results enhance the attractiveness of this idea by showing that features similar to zero-crossings in a filtered image can be rich in information. First, Ben Logan of Bell Laboratories has proved that a one-dimensional signal filtered through a certain class of filters can be reconstructed from its zero-crossings alone. The Laplacian of a Gaussian does not satisfy Logan's conditions exactly. Still, his work suggests that the primitive symbols provided by zero-crossings are potent visual symbols. More recently Alan Yuille and I have made a theoretical analysis of center-surround filtering. We have been able to show that zero-crossing maps obtained at different scales can represent the original image completely, that is, without any loss of information.

This is not to say that zero-crossings are the optimal coding scheme for a process such as stereopsis. Nor is it to insist that zero-crossings are the sole basis of biological vision. They are a candidate for an optimal coding scheme, and they (or something like them) may be important among the items to be matched between the two retinal images. We have, therefore, a possible answer to the question of what the stereop-

sis module matches. In addition we have the beginning of a computational theory that may eventually give mathematical precision to the vague concept of "edges" and connect it to known properties of biological vision, such as the prominence of "edge detector" cells discovered at the Harvard Medical School by David H. Hubel and Torsten N. Wiesel in the part of the cerebral cortex where visual data arrive.

To summarize, a combination of computational arguments and biological data suggests that an important first **step for stereopsis and other visual**

processes is the detection and marking of changes in intensity in an image at different spatial scales. One way to do it is to filter the image by the Laplacian of a Gaussian; the zero-crossings in the filtered array will then correspond to intensity edges in the image. Similar information is implicit in the activity of ON-center and OFF-center ganglion cells in the retina. To explicitly represent the zero-crossings (if indeed the brain does it at all) a class of edge-detector neurons in the brain (no doubt in the cerebral cortex) would have to perform specific operations on the output of ON-center

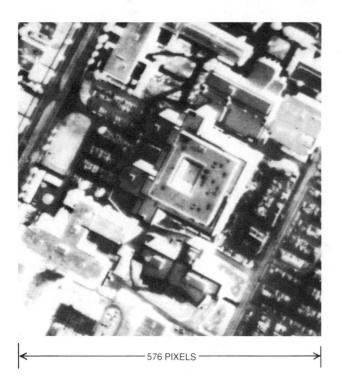

576 PIXELS

5 PIXELS

16 PIXELS

CENTER-SURROUND FILTERING of an image serves both to smooth it and to take its second spatial derivative. Here an image is shown at the left. Then filters of two sizes are shown. They are depicted schematically; the "filter" is actually computational. Specifically each intensity measurement in the image is replaced by a weighted average of neighboring measurements. Nearby measurements contribute positive weights to the average; thus the filter's center is "excitatory" (*red*). Then comes an annulus, or ring, in which the measurements contribute negative weights; thus the filter's "surround" is "inhibitory" (*blue*). The third part of the illustration shows the maps produced by the filters. They are no longer gray-level arrays. For one thing the maps have both positive values (*red*) and negative values (*blue*). They are maps of the second derivative. Transitions from one color to the other are zero-crossings; that is, they mark the places in the original image where its intensity changed most rapidly. The maps at the right of the illustration emphasize the zero-crossings by showing only positive regions (*red*) and negative regions (*blue*).

and OFF-center cells that are neighbors in the retina. Here, however, one comes up against the lack of information about precisely what elementary computations nerve cells can readily do.

We are now in a position to see how a representation of intensity changes might be useful for stereopsis. Consider first an algorithm devised by Marr and me that implements the constraints discussed above, namely uniqueness (a given point on a physical surface has only one location, so that only one binocular match is correct) and continuity (varia-tions in depth are generally smooth, so that binocular disparities tend to vary smoothly). It is successful at solving ran-dom-dot stereograms and at least some natural images. It is done by a comput-er; thus its actual execution amounts to a sequence of calculations. It can be thought of, however, as setting up a three-dimensional network of nodes, where the nodes represent all possible intersections of lines of sight from the eyes in the three-dimensional world. The uniqueness constraint will then be implemented by requiring that the nodes along a given line of sight inhibit one another. Meanwhile the continuity constraint will be implemented by re-quiring that each node excite its neigh-bors. In the case of random-dot stereo-grams the procedure will be relatively simple. There the matches for pixels on each horizontal row in one stereogram need be sought only along the corre-sponding row of the other stereogram.

The algorithm starts by assigning a value of 1 to all nodes representing a binocular match between two white pix-els or two black pixels in the pair of stereograms. The other nodes are giv-en a value of 0. The 1's thus mark all

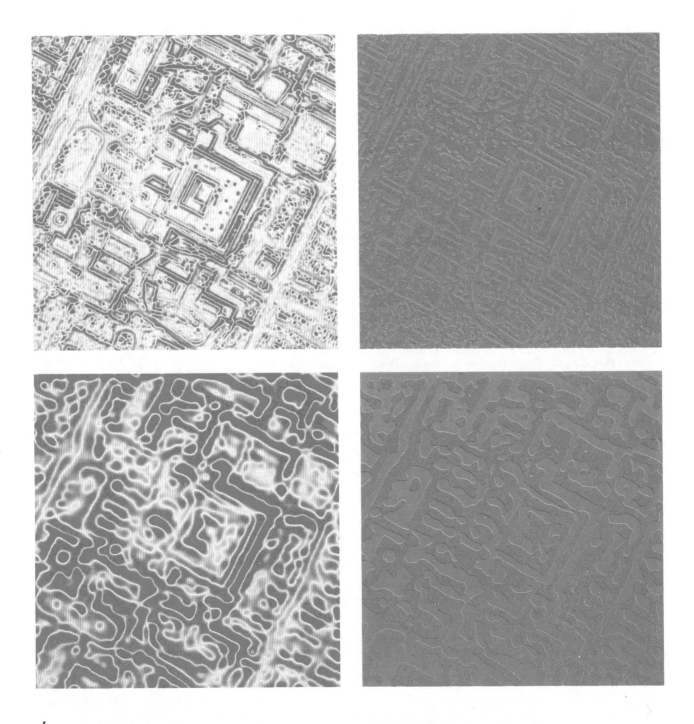

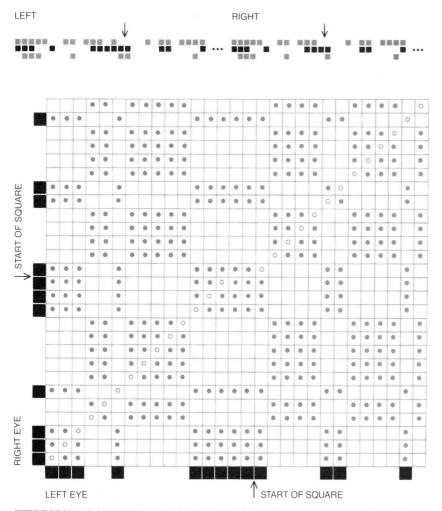

STEREOPSIS ALGORITHM devised by Marr and the author reconstructs the three-dimensional visual world by seeking matches between dots on corresponding rows of a pair of random-dot stereograms. At the top of the illustration two such rows are shown (*black and white*). Below them the rows are placed along the axes of a chart. Horizontal lines across the chart then represent lines of sight for the right eye; vertical lines represent lines of sight for the left eye. Color marks all intersections at which the eyes both see a black dot or a white dot. The problem is plain. A given black dot in one stereogram could in principle match any black dot in the other. The same is true for the white dots. Yet only some matches are correct (*open colored circles*), that is, only some matches reveal that a square of random-dot texture has a binocular disparity. The explanation of the algorithm continues in the illustration on the next page.

matches, true and false [*see illustration above on this page*]. Next the algorithm performs an algebraic sum for each node. In it the neighboring nodes with a value of 1 contribute positive weights; the nodes with a value of 1 along lines of sight contribute negative weights. If the result exceeds some threshold value, the node is given the value of 1; otherwise the node is set to 0. That constitutes one iteration of the procedure. After a few such iterations the network reaches stability. The stereopsis problem is solved.

The algorithm has some great virtues. It is a cooperative algorithm: it consists of local calculations that a large number of simple processors could perform asynchronously and in parallel. One imagines that neurons could do

them. In addition the algorithm can fill in gaps in the data. That is, it interpolates continuous surfaces. At the same time it allows for sharp discontinuities. On the other hand, the network it would require to process finely detailed natural images would have to be quite large, and most of the nodes in the network would be idle at any one time. Furthermore, intensity values are unsatisfactory for images more natural than random-dot stereograms.

The algorithm's effectiveness can be extended to at least some natural images by first filtering the images to obtain the sign of their convolution with the Laplacian of a Gaussian. The resulting binary maps then serve as inputs for the cooperative algorithm. The maps themselves are intriguing. In the ones gener-

ated by large filters at correspondingly low spatial resolution, zero-crossings of a given sign (for instance the crossings at which the sign of the convolution changes from positive to negative) turn out to be quite rare and are never close to each other. Thus false targets (matches between noncorresponding zero-crossings in a pair of stereograms) are essentially absent over a large range of disparities.

This suggests a different class of stereopsis algorithms. One such algorithm, developed recently for robots by Nishihara, matches positive or negative patches in filtered image pairs. Another algorithm, developed earlier by Marr and me, matches zero-crossings of the same sign in image pairs made by filters of three or more sizes. First the coarsely filtered images are matched and the binocular disparities are measured. The results are employed to approximately register the images. (Monocular features such as textures could also be used.) A similar matching process is then applied to the medium-filtered images. Finally the process is applied to the most finely filtered images. By that time the binocular disparities in the stereo pair are known in detail, and so the problem of stereopsis has been reduced to trigonometry.

A theoretical extension and computer implementation of our algorithm by W. Eric L. Grimson at the Artificial Intelligence Laboratory works quite well for a typical application of stereo systems: the analysis of aerial photographs. In addition it mimics many of the properties of human depth perception. For example, it performs successfully when one of the stereo images is out of focus. Yet there may also be subtle differences. Recent work by John Mayhew and John P. Frisby at the University of Sheffield and by Julesz at Bell Laboratories should clarify the matter.

What can one say about biological stereopsis? The algorithms I have described are still far from solving the correspondence problem as effectively as our own brain can. Yet they do suggest how the problem is solved. Meanwhile investigations of the cerebral cortex of the cat and of the cerebral cortex of the macaque monkey have shown that certain cortical neurons signal binocular disparities. And quite recently Gian F. Poggio of the Johns Hopkins University School of Medicine has found cortical neurons that signal the correct binocular disparity in random-dot stereograms in which there are many false matches. His discovery, together with our computational analysis of stereopsis, promises to yield insight into the brain mechanisms underlying depth perception.

One message should emerge clearly: the extent to which the computer and the brain can be brought together for the study of problems such as vision. On the one hand the computer provides a powerful tool for testing computational theories and algorithms. In the process it guides the design of neurophysiological experiments: it suggests what one should look for in the brain. The impetus this will give brain research in the coming decades is likely to be great.

The benefit is not entirely in that direction; computer science also stands to gain. Some computer scientists have maintained that the brain provides only existence proofs, that is, a living demonstration that a given problem has a solution. They are mistaken. The brain can do more: it can show how to seek solu-

tions. The brain is an information processor that has evolved over many millions of years to perform certain tasks superlatively well. If we regard it, with justified modesty, as an uncertain instrument, the reason is simply that we tend to be most conscious of the things it does least well—the recent things in evolutionary history, such as logic, mathematics and philosophy—and that we tend to be quite unconscious of its true powers, say in vision. It is in the latter domains that we have much to learn from the brain, and it is in these domains that we should judge our achievements in computer science and in robots. We may then begin to see what vast potential lies ahead.

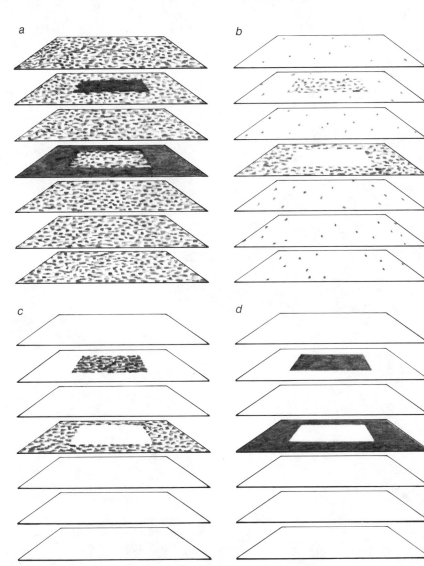

ITERATIONS OF THE ALGORITHM (depicted schematically) solve the problem of stereopsis. The algorithm assigns a value of 1 to all intersections of lines of sight marked by a match. The others are given a value of 0. Next the algorithm calculates a weighted sum for every intersection. Surfaces in the three-dimensional world tend to vary smoothly in depth; hence neighboring intersections with a value of 1 contribute positive weights to the sum. The eye sees only one surface along a given line of sight; hence intersections with a value of 1 along lines of sight contribute negative weights. If the result exceeds a threshold value, the intersection is reset to 1; otherwise it is reset to 0. After a few iterations of the procedure the calculation is complete: the stereograms are decoded. Natural images transformed into binary arrays (that is, into two-value zero-crossing maps) after center-surround filtering can be processed similarly.

9

The Hearing of the Barn Owl

by Eric I. Knudsen
December 1981

*The bird exploits differences between the sound
in its left and right ears to find mice in the dark.
It can localize sounds more accurately than
any other species that has been tested*

For the barn owl life depends on hearing. A nocturnal hunter, the bird must be able to find field mice solely by the rustling and squeaking sounds they make as they traverse runways in snow or grass. Like predators that hunt on the ground, the barn owl must be able to locate its prey quickly and precisely in the horizontal plane. Since the bird hunts from the air, it must also be able to determine its angle of elevation above the animal it is hunting. The owl has solved this problem very successfully: it can locate sounds in azimuth (the horizontal dimension) and elevation (the vertical dimension) better than any other animal whose hearing has been tested.

What accounts for this acuity? The answer lies in the owl's ability to utilize subtle differences between the sound in its left ear and that in its right. The ears are generally at slightly different distances from the source of a sound, so that sound waves reach them at slightly different times. The barn owl is particularly sensitive to these minute differences, exploiting them to determine the azimuth of the sound. In addition the sound is perceived as being somewhat louder by the ear that is closer to the source, and this difference offers further clues to horizontal location. For the barn owl the difference in loudness also helps to specify elevation because of an unusual asymmetry in the owl's ears. The right ear and its opening are directed slightly upward; the left ear and its opening are directed downward. For this reason the right ear is more sensitive to sounds from above and the left ear to sounds from below.

The differences in timing and loudness provide enough information for the bird to accurately locate sounds both horizontally and vertically. To be of service to the owl, however, the information must be organized and interpreted. Much of the processing is accomplished in brain centers near the beginning of the auditory pathway. From these centers nerve impulses travel to a network of neurons in the midbrain that are arranged in the form of a map of space. Each neuron in this network is excited only by sounds from one small region of space. From this structure impulses are relayed to the higher brain centers. The selection of sensory cues and their transformation into a map of space is what enables the barn owl to locate its prey in total darkness with deadly accuracy.

The barn owl has a wide range, both as a species and as an individual hunter. Barn owls are found throughout the tropical and temperate areas of the world. Many live close to human settlements, often nesting in barns or in belfries; they also nest in hollow trees and in holes in earth banks or rocks. Like most other owls they remain paired for long periods, sometimes for life, returning to breed in the same place year after year. The birds hunt in open areas, and they cover more ground than any other nocturnal bird. Studies of the bird's pellets (small objects coughed up by the bird that contain the indigestible remnants of prey) have shown that more than 95 percent of its prey are small mammals, mainly field mice; the rest are amphibians and other birds.

The nine species of barn owls are different enough from other owls to form their own family: the Tytonidae. The common barn owl, *Tyto alba,* is the most numerous species. It stands between 12 and 18 inches high and has a white face, a buff-colored back and a buff-on-white breast; its lower parts are mostly white with dark flecks. Each of the bird's middle toes has a small comb with which it dresses its feathers.

The most striking anatomical feature of the barn owl, and the one that plays the most important role in its location of prey, is the face. The skull is relatively narrow and small and the face is large and round, made up primarily of layers of stiff, dense feathers arrayed in tightly packed rows. The feathered structure, called the facial ruff, forms a surface that is a very efficient reflector of high-frequency sounds.

Two troughs run through the ruff from the forehead to the lower jaw, each about two centimeters wide and nine centimeters long. The troughs are similar in shape to the fleshy external pinna of the human ear, and they serve the same purpose: to collect high-frequency sounds from a large volume of space and funnel them into the ear canals. The troughs join below the beak but are separated above it by a thick ridge of feathers. The ear openings themselves are hidden under the preaural flaps: two flaps of skin that project to the side next to the eyes. The entire elaborate facial structure is hidden under a layer of particularly fine feathers that are acoustically transparent. The acoustic properties of the facial ruff are closely associated with the bird's method of locating the source of the sound.

In order to survive the barn owl must be able to locate prey with sound alone. Field mice are difficult to see even in broad daylight because their coloring blends with that of their surroundings; in addition they tend to travel through tunnels in grass or snow. By night, when the mice forage, they are essentially invisible even to the keen eyes of the owl. Hunting from the air makes the task even more difficult, since the owl must determine the angle of elevation above the prey. A determination of azimuth alone would leave an entire line of possi-

STRIKE OF THE BARN OWL in total darkness is shown in this sequence of exposures made in the laboratory with infrared radiation, to which the eyes of the owl are not sensitive. Unlike ground-living predators, the bird must determine its elevation above the prey as well as the direction in the horizontal plane. The owl can locate mice solely by sound. Moreover, just before striking it aligns its talons with the long axis of the mouse's body, as is shown in the final exposure. This action shows that the bird can infer the direction of the prey's motion from sound.

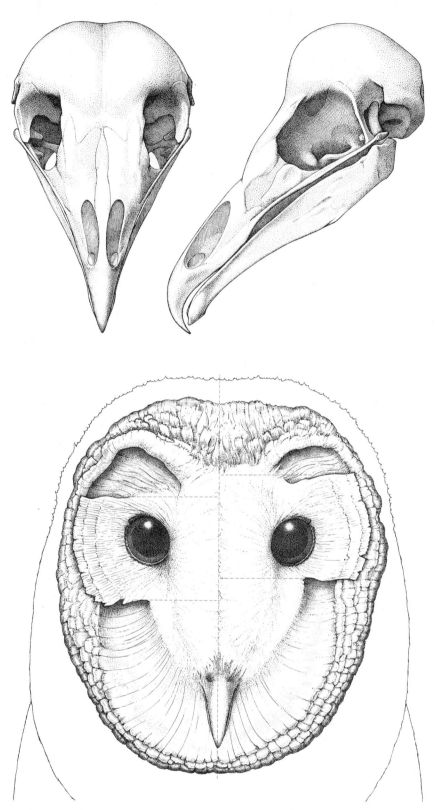

FACIAL STRUCTURE of the barn owl is responsible for much of the bird's precision in hearing. The face is formed by rows of tightly packed feathers called the facial ruff extending from the relatively narrow skull. The external ears are troughs formed by the ruff that run down the length of the face to join below the beak. They collect sounds and funnel them into ear-canal openings hidden under the preaural flaps (flaps of skin projecting outward from next to the eyes). The left ear is more sensitive to low-frequency sounds from the left and the right ear is more sensitive to those from the right. At high frequencies, however, the right ear is more sensitive upward, because the preaural flap and opening are lower on the right and the trough is tilted up. The opening and the flap are higher on the left side and the trough is tilted down. The left ear is more sensitive to sounds coming from below. Differences in perceived loudness can therefore yield clues to the elevation of a source of sound as well as to its horizontal direction.

ble target sites along the ground below.

The barn owl locates sounds in two spatial dimensions with great accuracy. Roger S. Payne, and later Masakazu Konishi and I, demonstrated that the bird is capable of locating the source of a sound within a range of one to two degrees in both azimuth and elevation; one degree is about the width of a little finger at arm's length. Surprisingly, until the barn owl was tested, man was the species with the greatest known ability to locate the source of a sound; human beings are about as accurate as the owl in azimuth but are three times worse in elevation. Monkeys and cats, other species with excellent hearing, are about four times worse than owls in locating sounds in the horizontal dimension, the only one in which they have been tested.

The sensitivity of the barn owl's hearing is shown both by its capacity to locate distant sounds and by its ability to orient its talons for the final strike. When the owl swoops down on a mouse, even in a completely dark experimental chamber, it quickly aligns its talons with the body axis of the mouse. It was Payne who first suggested that this behavior is not accidental. When the mouse turns and runs in a different direction, the owl realigns its talons accordingly. This behavior clearly increases the probability of a successful strike; it also implies that the owl not only identifies the location of the sound source with extreme accuracy but also detects subtle changes in the origin of the sound from which it infers the direction of movement of the prey.

Several kinds of experiments have helped to elucidate how the barn owl accomplishes these difficult tasks. Experiments with birds in free flight have measured how accurately the owl flies toward and strikes at an invisible sound source. Head-orientation experiments, where the bird was perched on a testing stand, have helped to measure the precision with which it aligns its head with an incoming sound. Experiments where the brain of an anesthetized owl was probed with a microelectrode while the bird was exposed to various sounds have shown how the sensory information is organized and interpreted by the central nervous system.

Konishi and I have done a series of head-orientation experiments. This experimental system has several advantages over tests conducted with birds in free flight. In free-flight tests flight errors may be confused with sound-location errors. In addition the angle of the sound source in relation to the bird's head at the moment the bird decides to strike cannot be determined. Free-flight trials are also complicated and time-consuming to conduct. In contrast, head-orientation experiments are relatively simple to conduct, and they allow the rela-

HEAD-ORIENTATION TESTS measured the accuracy of the owl's hearing. These experiments take advantage of a natural response made by the hunting owl: on hearing a noise the bird turns its head to face the source of sound. In the experimental situation the bird remained perched on a stand. A "search" coil mounted on its head was placed at the intersection of horizontal and vertical magnetic fields induced by stationary coils. Any head movement caused a measurable change in the current in the search coil. The owl's attention was first directed to a sound coming from a fixed "zeroing" speaker. A sound from a movable "target" speaker then caused the bird to turn its head in a rapid movement. A computer controlled the location of the target speaker and recorded movements of the bird's head. It was thereby possible to measure the accuracy with which the owl responded to sounds coming from various positions in space.

tion between the head and the sound source to be measured.

These experiments take advantage of a natural response of the owl when it is hunting. On hearing a noise the bird turns its head in a rapid flick that brings the source of the sound directly in front of it. This movement brings the sounds into the region where the bird's hearing is keenest. The eyes of the barn owl are immobile, and so the movement also enables the bird to see a target with maximum acuity. Konishi and I monitored this behavior by mounting a lightweight "search" coil on top of the owl's head. Magnetic fields generated by other coils were centered so that when the bird perched normally, the search coil was at the intersection of the horizontal field and the vertical one. The electric current induced in the search coil varied with its orientation to these fields. By evaluating the magnitude of two distinguishable signals one could measure the horizontal and vertical components of the orientation of the owl's head.

The tests were done in a totally dark chamber lined with materials that eliminate echoes. Sounds were generated by a stationary speaker (the "zeroing" speaker) and a movable speaker (the "target" speaker). The owl first turned to face a sound from the zeroing speaker, placed in front of its perch; a sound delivered by the target speaker then caused the bird to turn its head in the characteristic flicking movement. A computer controlled the location of the target speaker and recorded its location and the alignment of the owl's head.

The head-orientation trials have yielded much information about how the owl determines the origin of a sound. One of the important features of the process is that it can achieve maximum accuracy even with sounds that end before the head movement begins. This indicates that the owl's auditory system determines the azimuth and elevation of a sound without head movement and then utilizes the information to direct the head-orientation response. The head-movement tests also show that the owl's accuracy deteriorates with increases in the angle between the source of the sound and the orientation of the bird's head.

Our experiments and those of others have shown that the barn owl's ability to locate the origin of a sound is dependent on the presence of high frequencies in the sound. Although the owl's hearing is sensitive to a broad range of frequencies, from 100 hertz (cycles per second) to 12,000 hertz, it can locate accurately only sounds with frequencies between 3,000 and 9,000 hertz. In addition experiments in which one of the bird's ears is plugged show that both ears are necessary for the accurate locating of targets. If one ear is plugged, the owl makes large errors in the direction of that ear.

With these characteristics in mind we proceeded to investigate the exact information the barn owl selects from natural sounds. To locate the source of a sound the owl must determine the direction of propagation of the sound waves based on information from detectors at two points, namely its ears. The most useful spatial information is gained by comparing information from these two sources, since the differences between them depend not on the absolute sound level but only on the orientation of the ears in the sound field.

One valuable cue of this kind is the difference in the time of arrival of the sound in the two ears. When the sound comes directly from the side, the difference is at its maximum; when the sound is directly in front of the bird, there is no difference in the arrival time at the two ears. Between these limits the time difference varies with the angle of the sound in the horizontal plane. The time delay can therefore yield information about the azimuth of the sound. This is not sufficient for the bird to locate the sound exactly, because sounds from several directions can give rise to the same difference in time. In three-dimensional space these directions form a cone around the axis between the owl's ears.

This is an example of the inadequacy of any one cue to provide information about both the horizontal and the vertical angle of a source of sound. To specify a location in both dimensions two independent cues are needed. In the owl's case the additional information is provided by differences in the directional sensitivity of the ears.

Directional sensitivity is provided by the facial ruff. Like a hand cupped behind the ear, the troughs of the ruff amplify the sound and make the ear more sensitive to sounds from certain direc-

tions. The amount of amplification and directional sensitivity imparted by the feathers of the facial ruff varies dramatically with the frequency of the sound. This is owing to one of the properties of the sound waves themselves. When sound waves encounter an object, they can bend around it or be reflected back from it. Which of these happens depends on the wavelength of the sound and the size of the object. If the wavelength is long compared with the object, the waves tend to propagate around the object; if the wavelength is short, the waves tend to be reflected back in the direction from which they came.

As a result of this phenomenon frequencies of less than 3,000 hertz are not efficiently reflected by the ruff. Because the funneling action of the ear depends on its capacity to reflect sound its directional sensitivity at low frequen-

cies is relatively poor. At 3,000 hertz, for example, the left ear is only slightly more sensitive to sounds coming from an area between 20 and 40 degrees to the left than it is to sounds coming from other directions. The right ear has a similar degree of sensitivity to the right. Since the sensitivity of each ear at low frequencies changes only gradually with direction, the comparison of sound intensities at low frequencies can provide only a coarse spatial cue. Moreover, this difference yields no clue to the elevation of the sound.

With higher-frequency sound waves the situation is quite different. Each ear is much more sensitive to the direction of the sound; a small change in sound direction gives rise to a large change in perceived intensity. In addition, instead of being more sensitive to the right or to the left, the right ear is more sensitive above the horizontal plane and the left

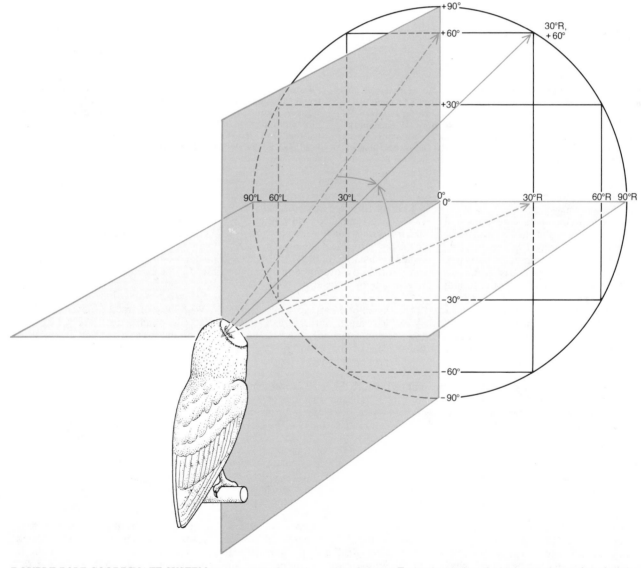

DOUBLE-POLE COORDINATE SYSTEM serves to map the positions of the owl's head in head-orientation tests. Each direction is specified by two measurements: the angle from the horizontal and that from the vertical through which the owl must turn its head to face that way. Facing forward, as the bird normally perches, the orientation is 0 degrees (horizontal) and 0 degrees (vertical). The map represents the 180 degrees of space in front of the bird. The other maps accompanying this article have the same coordinate system.

ear is more sensitive below it. This sensitivity is the result of an unusual anatomical asymmetry. The ruff on the left is directed slightly downward, and the ear opening and the preaural flap are higher in the ruff on the left side. On the right side the reverse is the case. Accordingly as the source of the sound moves up the high-frequency components of a natural sound become louder in the right ear and softer in the left. As the source of the sound moves down the sounds become louder in the left ear. Since at high frequencies the perceived loudness changes rapidly with elevation, this cue offers information that is very precise.

Although that information is valuable to the barn owl, it is also complex. The magnitude of the difference in intensity varies according to the frequency of the sound, because of the greater capacity of the ruff to reflect sounds at higher frequencies. The direction indicated also varies with frequency, since horizontal location is given by low frequencies and vertical location by high ones. The owl's auditory system must therefore compare the intensities detected at each ear for each frequency. A comparison of intensities made frequency by frequency is called an interaural spectrum.

Since low-frequency sounds yield clues to azimuth and those of high frequency yield clues to elevation, the interaural spectrum could by itself provide enough information for the owl to locate prey. Much evidence supports the hypothesis that owls use this spectrum. The strike accuracy of the bird increases sharply as the bandwidth (the number of frequencies contained in a sound) is increased. From differences in the intensity of a single tone (a sound of only a single frequency) the owl can determine direction in only one dimension; as the spectrum broadens more intensity differences are available, and their values indicate the angle of the source in more than one plane. Experiments in which one of the owl's ears is plugged show more directly that the owl compares intensities. A trained owl with its right ear plugged strikes to the left and short of the target; one with its left ear plugged strikes to the right and beyond the target. That the errors have components of both elevation and azimuth implies that the owl gains both types of information from comparisons of intensity.

Further confirmation has been obtained by removing the owl's facial ruff. When the ruff is removed, the owl is able to locate the azimuth of a sound quite well but cannot identify its vertical location: the bird consistently orients to a point on the horizontal, regardless of the elevation of the target. This accords with our hypothesis: the sound-reflecting properties of the ruff underlie directional sensitivity at high frequencies, which enables the owl to identify the

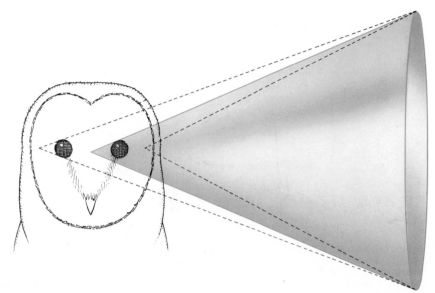

CONE OF CONFUSION (*color*) is formed by the directions among which the barn owl cannot distinguish on the basis of time delay alone. Time delay can provide information about the horizontal angle of a sound source. Since the left and right ears are generally at slightly different distances from the source of a sound, sound waves reach them at slightly different times. The greater the angle of a sound source from the bird's frontal plane, the greater the time delay. There are many directions, however, that give rise to the same path lengths (*broken lines*) and hence to the same time delay; other cues are required for the bird to tell them apart. In three-dimensional space these directions form a cone whose peak is between the two ears.

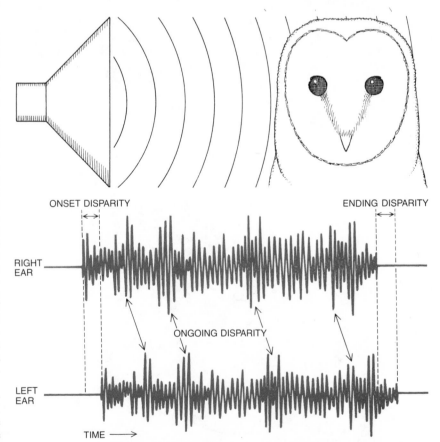

SOUND TRACE shows two kinds of differences in the timing of the sound between the barn owl's left and right ears. In this schematization the height of the trace indicates the pressure caused by the sound waves in each ear. When a sound comes from the right, the waves begin and end earlier in the right ear; major changes in intensity also occur slightly earlier there. These differences are collectively known as transient disparity. In addition, throughout the duration of the sound the waveforms are slightly advanced in the right ear; this is called ongoing disparity. Man relies on both differences to find the source of a sound. The barn owl relies heavily on ongoing disparity, but there is no evidence that the bird exploits the transient difference.

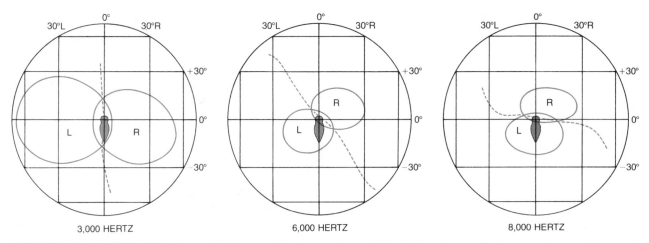

3,000 HERTZ   6,000 HERTZ   8,000 HERTZ

DIRECTIONAL SENSITIVITY of the ears of the barn owl changes with the frequency of sound waves. The funneling action of the troughs in the facial ruff makes the ears more sensitive to sounds from some directions than to sounds from others. At low frequency the area of maximum sensitivity of the left ear is to the left and that of the right ear is to the right. Waves with higher frequencies are more efficiently reflected by the ruff. Since the left trough is tilted down and the right trough up, as the frequency increases the areas of greatest sensitivity move toward vertical alignment: the right ear is more sensitive upward and the left ear downward. At 8,000 hertz (cycles per second) the regions of most sensitive hearing are almost directly above and below the bird. Because natural sound usually has many frequencies, the owl can exploit differences in loudness to identify both the horizontal and the vertical direction of a sound source.

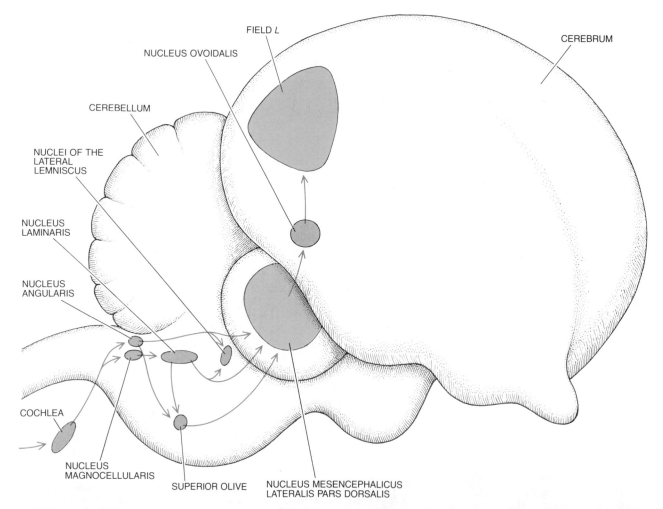

AUDITORY PATHWAY of the central nervous system of the barn owl leads from the cochlea to an area of the forebrain known as Field *L*. Along this route the nerve impulses undergo considerable processing: data about the timing and frequency of sound are converted into information about the location of the sound source. Much processing takes place in various centers of the lower brain. By the time the impulses have reached the nucleus of the midbrain that is called the mesencephalicus lateralis pars dorsalis (MLD) they are directed into a network of neurons that respond to sounds from specific areas; the distribution of those areas forms a two-dimensional map of the space in front of the bird. Information about a sound's location then passes to Field *L*, corresponding to the auditory cortex of mammals.

vertical angle of the source. Removal of the ruff eliminates the ability to discriminate among elevations. Some disparity in intensities is retained because the ear openings and preaural flaps are placed asymmetrically in the fold of skin supporting the feathers of the ruff, but this is not enough to make it possible to identify the elevation of the sound.

That barn owls rely on the interaural spectrum to locate sounds in both azimuth and elevation has thus been amply confirmed by the results gathered from the head-orientation tests. It is clear from other findings, however, that the bird also makes some use of timing differences in locating its invisible prey. The timing delay is manifested in two aspects of the binaural signal. First, the sound begins and ends sooner in the ear closer to the source; the timing of major discontinuities in intensity in the sound is also slightly different in each ear. These differences are known collectively as transient disparity. Second, throughout the duration of the sound the sound waves reaching the far ear will be slightly delayed. With a single frequency this difference in the timing of the waveforms is known as phase delay; with more complex natural sounds, made up of many frequencies, it is called ongoing time disparity.

In nature the ongoing and the transient disparities are of about equal magnitude; they vary with changes in the azimuth of the source of sound. They do, however, have different advantages for locating sounds. The ongoing time disparity can be measured repeatedly while the sound lasts. Transient disparity, on the other hand, can be monitored only intermittently, but it is less likely than the ongoing disparity to be confused by echoes.

Human beings rely on both transient and ongoing disparity to determine the source of a sound. Owls appear to rely on only the ongoing difference. Like other kinds of spatial information, the ongoing disparity has a major ambiguity. It may be best understood in the case of a tone coming directly from the side. The signals detected by the owl's ears are sinusoidal (regular) waves; because of the different distances to the ears the waves will be slightly out of phase with each other. The magnitude of the phase delay so created will depend both on the frequency of the tone and on the distance between the ears. As the frequency of the wave or the distance between the ears increases, the wave passes through more of its cycle as it travels around the head to reach the far ear; hence the phase delay is greater.

When the frequency of the tone is so high that the wave passes through exactly half of its cycle before it reaches the far ear, the phase delay corresponds to half of the wavelength. Such a delay could be caused by a sound coming directly from the owl's left or directly from its right, since the difference in path lengths is the same for these two directions. It is therefore impossible for the bird's auditory system to determine the direction of the sound on the basis of ongoing disparity alone. At higher frequencies the situation is worse still. When the wavelength is equal to the distance between the ears, for example, there is no phase delay, since the wave travels through its entire cycle while passing around the head. This relation could correspond to a sound coming from the right, from the left or from directly ahead; on the basis of phase delay alone the bird has no way of determining which is the case.

The wavelengths at which such ambiguities arise depend on the distance between the ears. The barn owl's ears are about five centimeters apart; phase ambiguity will therefore arise at a wavelength of 10 centimeters or less, which corresponds to frequencies of 3,000 hertz or more. Since the barn owl has no difficulty determining the azimuth of even high-frequency tones, Konishi and I assumed that at high frequencies the bird must rely on some source of information other than the ongoing time disparity. A likely candidate for the additional cue was transient disparity, because it is not affected by changes in frequency and because other species, including man, are known to depend on it at high frequencies.

How wrong this conclusion was has been demonstrated conclusively by Andrew Moiseff and Konishi in a further head-orientation experiment with the barn owl. They presented sound directly and independently to both ears by means of small speakers implanted in the owl's ear canals. This technique enabled them to eliminate transient disparities and differences in intensity between the ears as they varied the ongoing time disparity. The delay between the waveforms of the sound in the two ears could be adjusted in steps as small as one microsecond. In response to ongoing disparities of as little as 10 microseconds or as much as 80 microseconds the owl made quick horizontal turns of the head that corresponded approximately to the angle implied in the ongoing difference. This response suggests, in the absence of other interaural cues, that the owl continues to make use of the phase delay, or ongoing disparity, even at high frequencies.

The finding was startling because it implied that the barn owl has some means of overcoming the phase ambiguity. Furthermore, for the owl's auditory system to sense such small differences in the timing of the waveforms in the two ears it must receive specific information about acoustic signals occurring at 7,000 hertz, or once every 143 microseconds. This is remarkable, because the nerve impulses that convey this information from the cochlea to the brain last for more than 1,000 microseconds.

Ongoing time disparity is very useful to the barn owl; by itself, however, the cue does not have the precision the owl needs in order to hunt. With speakers implanted in the ear canals the owl responded to a given ongoing time disparity with turns that varied in magnitude by up to 15 degrees in either direction. In contrast, the largest standard deviation of error for owls responding to external targets is only 2.5 degrees in either direction, and the error is usually less than 1.5 degrees. Other cues must therefore be combined with the ongoing disparity. It is known that differences in intensity help to specify location, but we also tried to discover whether the bird employs transient disparities.

Our original hypothesis, that the owl relies on transient disparity to compensate for phase ambiguity at high frequencies, has clearly been invalidated. The owl's performance in determining the azimuth of a tone suggests that transient disparities are not relied on at all. In head-orientation trials the owl was presented with tones of 7,000 and 8,000 hertz. The speaker was sometimes at 30 degrees to the right and sometimes at 30 degrees to the left. In these situations a curious kind of behavior was observed: with the target at 30 degrees to the right the owl sometimes turned 30 degrees to the left, and vice versa.

This confusion is clearly a manifestation of phase ambiguity. The cycle period of a 7,000-hertz tone is 143 microseconds and that of an 8,000-hertz tone is 125 microseconds. With a source of sound at 30 degrees to the right or the left these tones pass through about half a cycle in traveling over the difference between the path lengths to the ears. Therefore a tone coming from 30 degrees to the right yields the same phase delay as one coming from 30 degrees to the left. The ambiguity is present only in the ongoing disparity; if the owl had been relying on transient disparity, it would immediately have picked out the correct location.

This result confirms that the owl continues to rely on phase-delay information, even at high frequencies, in spite of its ambiguity, and apparently does not rely on transient disparity. Since the wavelength of low-frequency sounds is considerably greater than the distance between the ears, phase ambiguity does not exist at low frequencies. Moreover, the presence of a number of frequencies in natural sounds helps the owl to resolve phase ambiguities. When many different pairs of directions arise from different frequencies, the owl selects the one direction in space that is consistent with one member of each pair.

Head-orientation trials clearly show that the barn owl relies on two kinds of information to determine the origin of

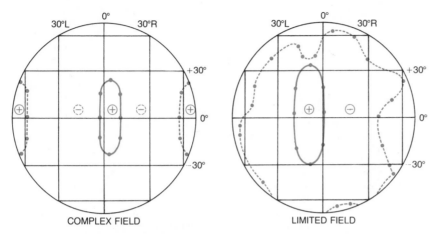

COMPLEX FIELD   LIMITED FIELD

TWO TYPES OF NEURONS in the brain of the barn owl are sensitive to sounds from specific directions. Shown here are maps of the receptive fields of the two cell types. Receptive fields are the regions of space within which sounds will produce an excitatory response in the neuron. Complex-field neurons, found in the MLD, have several such areas. A typical complex-field receptive pattern is shown at the left. The receptive areas correspond to the directions that give rise to identical ongoing disparities in the timing of sound waves. Between those regions sounds produce an inhibitory response. Limited-field neurons are found in both the MLD and Field *L*. A typical limited-field receptive pattern is shown at the right. Limited-field cells have a single receptive area. Sounds from outside this area produce a strong inhibitory response. Each limited-field cell responds to a specific difference between the ears in sound timing and intensity.

a sound: the interaural spectrum and the ongoing time difference. The latter yields clues to the azimuth of the sound source. The differences in intensity between the ears yield clues to both the azimuth and the elevation. To learn how the auditory system transforms these cues into a neural image of sound location calls for a different experimental approach.

The technique I resorted to, first with Konishi and later in independent experiments, is that of inserting a microelectrode into the brain of an anesthetized owl and searching for sites of neuronal activity while sounds are presented to the bird's ears. Konishi and I began the experiments with the same apparatus we had used in the head-orientation trials. After the owl was anesthetized its head was held rigidly in a special stereotaxic frame. A microelectrode was lowered into the brain until nerve impulses from a single neuron could be recorded. By moving the target speaker around the owl it was possible to map the regions of space to which the neuron responded.

Since the source of a sound in space is determined only after considerable neural processing, we began our study by exploring structures fairly far along the auditory pathway: in the midbrain and forebrain. The main auditory center in the midbrain of birds is called the nucleus mesencephalicus lateralis pars dorsalis (MLD). (It corresponds to the structure in the brain of mammals called the inferior colliculus.) Nerve impulses reaching this center have already been processed in one or more nuclei farther down the auditory pathway. Farther up the pathway an area designated Field *L* is the primary receiving center in the forebrain for auditory im-

pulses. (This structure corresponds to the auditory cortex in mammals; birds have no exact analogue to the auditory cortex.)

In both the MLD and Field *L* the large majority of neurons do not respond precisely to spatial cues. Some of the neurons show their highest level of activity in response to sounds from a certain region of space, but the borders of the region are not sharp, and they vary greatly with the intensity of a sound. Other neurons are even less specific in their response, being excited by sounds from virtually all directions. These neurons probably contribute not to specifying location but to the detection or identification of sounds.

Two types of neuron found in the MLD and Field *L*, however, are highly sensitive to sound location. The first of them, called the complex-field neuron, is found in large numbers in the MLD, usually in clusters scattered among other kinds of neurons. The activity of complex-field neurons is stimulated by sounds coming from several separate regions of space, called excitatory fields. The neurons are much less excited, or are even inhibited, by sounds arising in the regions of space between the excitatory fields.

When the location of the centers of the excitatory fields in space are calculated for an individual complex-field neuron, an interesting correspondence is observed. The excitatory fields are the same as the regions of space the owl confuses under conditions of phase ambiguity. The multiple excitatory fields represent the regions from which sounds reaching the ears will generate phase delays of equivalent magnitude. Each complex-field neuron therefore seems to be sensitive to a particular

phase delay at a particular frequency. The presence of several excitatory fields for each neuron would appear to be the physical correlate of spatial confusion due to phase ambiguity.

The second type of neuron, called the limited-field neuron, which is found in both the MLD and Field *L*, responds in an even more specific way. Limited-field cells are excited only by sounds coming from a single region of space. The regions to which limited-field neurons respond are typically elliptical. Their size varies in azimuth from seven degrees to 42 degrees and in elevation from 23 degrees to an entire band in front of the bird. Unlike other neurons in the auditory pathway, the limited-field neurons are extremely selective, responding only to changes in location; large changes in sound intensity cause little if any alteration in the sharp borders of the receptive region.

Some of the sharpness of the borders of the limited-field-neuron receptive regions is due to the fact that sounds coming from outside the excitatory region inhibit the cell's response. The inhibitory effect becomes stronger as the location of the sound approaches the border of the excitatory region, at which point the inhibiting effect changes to one of excitation. Although the spatial regions that give rise to an excitatory response are fairly large, within them there is a smaller region of the best response. These "best regions" vary in size from 2.5 to 15 degrees in azimuth and from five to 45 degrees in elevation.

Limited-field units are found in both the MLD and Field *L*. Their distribution in the two centers is, however, fundamentally different. In Field *L* limited-field neurons make up only about 15 percent of the total population of nerve cells, and they are often found scattered among other types. In the MLD, on the other hand, these cells are concentrated on the side and front margins of the nucleus, interspersed only with a few neurons of the complex-field type. After we had intensively explored and mapped the MLD it became apparent to us that the arrangement of limited-field neurons in that nucleus constitutes a map of two-dimensional space, with the distribution of the receptive areas of the neurons following the contours of space.

The map is distorted, however: the area in front of the bird, the region of maximum auditory acuity, is represented disproportionately. Sound azimuths are arrayed in the horizontal plane. On the right side of the structure representing the map are neurons that are excited by sounds originating between 15 degrees to the right and 60 degrees to the left. On the opposite side are neurons stimulated by sounds between 15 degrees to the left and 60 degrees to the right. This arrangement means that the 30 degrees of space in front of the bird is represented on both sides of the map and therefore by two sets of neurons. In addition the map is arranged so that on both sides the population of neurons that represent the 30 degrees in front is disproportionately large. As a result of the double representation and the large number of neurons on each side the 30 degrees in front is analyzed with great precision, a fact that may explain the owl's particular accuracy in locating sounds in this area.

Sound elevations are arrayed transversely on the map. The "best regions" of the limited-field neurons range from 40 degrees upward to 80 degrees downward. The upper fields are at the top of the curved surface of the map and the lower fields are at the bottom.

How does the nervous system of the barn owl construct this remarkable map? Substantial understanding of the process has come from experiments by Moiseff and Konishi. In these tests speakers were placed in the ear canals of anesthetized owls. When sound was delivered separately to the two ears, the timing and intensity differences required to elicit a response in each neuron could be observed. These values were then correlated with the areas of space known to excite each set of neurons. The results of the investigations show that the limited-field units in the map are quite sensitive to ongoing differences in timing. This observation corresponds nicely to the bird's dependence on ongoing disparities demonstrated in head-orientation experiments.

Limited-field neurons respond only to an extremely narrow band of ongoing time delays: the size of the band ranges from 40 to 100 microseconds. Even within this minute range one particular delay always elicited the greatest response. Changing the ongoing difference by as little as 10 microseconds could change the strength of the neuron's response by as much as 75 percent. This degree of sensitivity complements (and helps to explain) the owl's precision in aerial hunting.

The ongoing disparity that gives rise to the greatest response also corresponds to the region of space to which the cell responds. Neurons that respond to sounds coming from the front are maximally excited by small ongoing disparities. Neurons that respond to sounds at greater angles require larger disparities. These results confirm at the level of individual cells the conclusion drawn from behavioral tests: that the owl relies heavily on the ongoing difference in timing between the ears to determine location in the horizontal plane.

Such experiments have also helped to confirm and explain the owl's reliance on differences in sound intensity. By varying the relative intensity of sounds presented independently to the ears of an anesthetized owl it was found

that for each limited-field neuron there is one difference in intensity that evokes the maximum response. Changing the difference away from this value causes the response to decrease and finally to cease. The pattern is not affected by the changes in the average sound level; it depends only on the difference in intensity between the ears.

The evidence derived from probing the brain of the barn owl shows that the map of space in the MLD is created by the same cues the owl was known to rely on in finding the azimuth and eleva-

tion of a sound: ongoing time disparities and the interaural spectrum. By means of an exquisitely precise transformation the auditory system converts these cues into spatial information. The arrangement of the cells of the map implies that neighboring neurons respond to cues that are only very slightly different. Moreover, the order of the excitatory fields in the map follows the continuity of space. How the brain of the owl is able to achieve such precise connections presents an intriguing problem for further investigation.

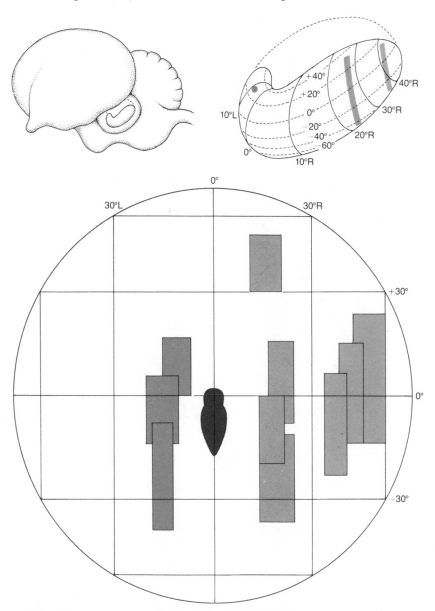

TWO-DIMENSIONAL MAP of frontal space is found in the MLD of the barn owl. The map is made up of limited-field neurons. Each cell responds to sounds from a specific region of space. The arrangement of the receptive areas of the neurons follows spatial contours. Shown above is the region on the left side of the brain that responds to sounds from the right. Neurons in this region are sensitive to sounds in an area from 15 degrees on the left to 60 degrees on the right, and from 40 degrees upward to 80 degrees downward. Electrode traces through the structure will thus correspond to a set of regions of space aligned vertically; a set of three such traces and the spatial regions they correspond to are shown. A corresponding structure on the right side of the owl's brain consists of the neurons that respond to sounds originating from the left. The region from 15 degrees on the left to 15 degrees on the right is thus represented on both sides of the brain and is mapped on each side by a disproportionately large number of neurons; this is the area of the barn owl's greatest acuity in locating a sound source.

V

BRAIN, BEHAVIOR, AND MIND

V BRAIN, BEHAVIOR, AND MIND

INTRODUCTION

I n the last analysis, the brain has only one function—behavior. The brain evolved and expanded because it provided for more adaptive behavior. A larger brain evidently increases the options for survival. This final section includes articles representing three areas within the broad field of behavioral neuroscience: behavioral genetics, mental illness, and the mind-body problem.

The "nature-nurture" question has been a large issue in biology and psychology. Do our genes or our environment determine what we are? Is intelligence determined genetically or experientially? Such questions are virtually meaningless, as are similar questions about the complex traits and abilities of higher animals and humans, because from at least the time the ovum is fertilized, the developing organism is influenced by both. The rapidly developing field of behavioral genetics contradicts the nature versus nurture idea.

Life begins with the genes. The more we learn about the gene, the more complicated the process seems. Genes have been characterized, recombined, and engineered in astonishing ways by molecular biologists. More recent discoveries—such as the ability of certain genes in corn to move about on chromosomes, to say nothing of the complex process that leads from genes to cells to organisms—clearly demonstrate that the relationship between genes and behavior is not simple.

Certain behavioral patterns, however, particularly in simpler animals, are relatively unvarying and only minimally influenced by experience. The wiring diagrams among the relatively few neurons they possess are fixed by the genes. In the first article of this section, Richard Scheller of Stanford University and Richard Axel of Columbia University describe a striking example of how genes regulate egg-laying behavior in the *Aplysia,* an invertebrate with a relatively simple nervous system.

Given the staggering complexities of the human mind and its disorders, psychiatry, psychology, and neuroscience have made surprising progress in the past 30 years in understanding and treating mental illness. Depression is a condition everyone has experienced to varying degrees. Normal people become sad and depressed for good reasons such as the loss of a loved one or a financial setback, etc. In psychotic depression, however, the reason, or at least an appropriate reason, is not apparent externally. The symptoms of psychotic depression are like those of normal depression but are exaggerated. Unlike schizophrenia, thought processes remain relatively normal.

There are actually two different forms of depression: major depression, just described, and bipolar depression, formerly called manic depression. The diagnosis of manic depression depends on at least one episode of mania, which is characterized by unreasonable euphoria and feelings of intense joy and power that result in bizarre manic behavior. Manic episodes are quite rare; the depressed phase is much more common. Although major and bipolar depression seem similar, treatment for the two conditions differs. Lithium, in particular, has been a great help in the treatment of bipolar depression. In the next article, Daniel Rosteston of the Harvard Medical School explores the possible neuronal-cellular basis of the effect of lithium on mania.

In the last article, Jerry Fodor of the Massachusetts Institute of Technology addresses the ultimate question: The mind-body problem—the nature of the mind. He introduces the major philosophic positions regarding the nature of the mind and presents a modern, functionalist view that has developed from the interaction of such fields as computer science, artificial intelligence, linguistics, and psychology.

10

How Genes Control
an Innate Behavior

by Richard H. Scheller and Richard Axel
March 1984

The techniques of recombinant DNA are exploited to define a family of genes encoding a set of related neuropeptides whose coordinated release governs a fixed-action pattern: egglaying in a marine snail

Certain stereotyped patterns of animal behavior are innate. They are shaped by evolution and are inherited by successive generations; largely unmodified by experience or learning, they are displayed by all individuals of a species and not by other species. Ethologists have described such innate, stereotyped behavioral arrays as "fixed-action patterns": patterns of behavior consisting of several independent elements that either together form a coordinated sequence or do not take place at all. Each animal inherits a unique collection of such fixed-action patterns, which characterizes the behavior of its species.

How does an animal inherit such a behavioral repertoire? What does an animal inherit? It inherits DNA. The genes of the DNA can specify stereotyped behavior in two ways. First of all, they can specify a precise network of interconnected nerve and muscle cells that are put in place and "wired" together in the course of the animal's development. A stereotyped behavior is elicited, however, only in particular situations or at particular stages of an animal's life cycle, and only by the coordinated activity of particular parts of the network of neurons and muscle cells. In addition to the network, then, the genes must specify control elements: substances that excite specific preexisting connections in a rigidly determined way to generate a fixed-action pattern at the right time.

As molecular biologists studying the nervous system we seek to identify such control elements and the genes governing their synthesis. In so doing we hope eventually to learn something about the factors that play a role in the generation of innate behavioral repertoires, how they evolve in different species and how they develop in a given species.

Specific behaviors, however, unlike such traits as eye color or the inherited disorders of hemoglobin, are not likely to directly reflect the state of specific genes. The central nervous system integrates and filters the dictates of genes in ways that for the most part are inaccessible to experiment. The more complex the central nervous system, the more elusive the relation between a set of genes (a genotype) and observable traits (a phenotype). It is easier to study genes specifying behavior in an organism that is sophisticated enough to exhibit interesting behavioral repertoires but simple enough so that the behavior can be attributed to identifiable cells. We work with such an animal: the mollusk *Aplysia,* a shell-less marine snail that can weigh as much as five to 10 pounds.

The snail's central nervous system is numerically simple, consisting of about 20,000 nerve cells collected into four pairs of head ganglia and one abdominal ganglion. This is in striking contrast to the brain of a mammal, which has perhaps a million times as many neurons. Moreover, the neurons of *Aplysia* can be as much as a millimeter in diameter, more than 1,000 times the size of a typical human brain cell. Most of these huge cells contain correspondingly large quantities of DNA: as much as a microgram, or several hundred thousand times the DNA content of a typical mammalian neuron. Functions that would be carried out by a large collection of related neurons in a more complex nervous system may be handled in *Aplysia* by a single large cell. A series of investigations, notably the elegant work of Eric R. Kandel and his colleagues at the Columbia University College of Physicians and Surgeons, has related specific patterns of behavior in the snail to the functioning of particular cells. We and our colleagues have gone on to examine the activity of specific genes in a single cell and have attributed a behavior pattern to the activity of individual genes.

An adult *Aplysia* is largely occupied with feeding and reproduction. Aspects of the snail's reproductive behavior are highly ritualistic, involving a coordinated series of stereotyped patterns that accomplish courtship, mating and the deposition of fertilized eggs. *Aplysia* is a true hermaphrodite, an individual serving as both male and female, in most cases simultaneously, with the snails copulating in long chains of half a dozen animals or more; sometimes the chain is closed to form a circle. Fertilization takes place internally, in the reproductive duct, and then the fertilized eggs are laid to develop externally, in the sea. The genes we study are those implicated in the elaborate but completely stereotyped array of coordinated behaviors that accomplish the egg-laying process.

The eggs are laid in long strings of more than a million eggs. As the string of eggs is expelled by contraction of the reproductive-duct muscles, the snail stops walking. It stops eating. Its heart rate and respiratory rate increase. The snail grasps the egg string in its mouth. With a series of characteristic head-waving movements it helps to extract the string from the duct and winds the string into an irregular mass. A small gland in the mouth secretes a sticky mucoid substance that becomes attached to the mass. Then, with one forceful wave of the head, the animal affixes the entire mass of eggs to a solid support such as a rock. A number of disparate actions have come together in a rigidly coordinated sequence to serve a common function: the deposition of fertilized eggs in a way that will afford protection during their development.

What substances control this coordinated pattern of behavior and what genes encode these substances? First, where are the substances synthesized? Several years ago Kandel and Irving Kupfermann identified two clusters of neurons, the bag cells, at the top of the abdominal ganglion. When an extract of these cells was injected into live snails, it elicited the full repertoire of behaviors associated with egg laying even though the animals had not mated and the eggs were unfertilized. In subsequent studies Stephen W. Arch of Reed College and Felix Strumwasser of the California Institute of Technology isolated one of the active bag-cell factors and identified it as a peptide, or small protein, consisting of 36 amino acids. They found that when the peptide was administered to snails, it elicited some but not all of the egg-laying behaviors, suggesting it was one of several factors controlling the total behavioral repertoire. Strumwasser determined the linear sequence of the amino acids in the peptide, which was designated egg-laying hormone (ELH). Because there is a linear relation between the amino acid sequence of a pro-

tein and the nucleotide sequence of the gene encoding it, the identification and sequencing of a peptide controlling behavior put a problem in behavioral biology well within the realm of molecular genetics.

In collaboration with Linda B. McAllister, James F. Jackson, James H. Schwartz and Kandel we set out to isolate the gene encoding ELH from the *Aplysia* genome: the total complement of DNA in the snail's chromosomes. The procedure for isolating specific genes depends on the techniques of recombinant DNA. In brief, one establishes a "library" of recombinant-DNA molecules, each of them carrying a

small fragment of the *Aplysia* genome, and then scans the library with a probe that will detect the ELH gene. We cleaved the snail DNA into many thousands of fragments, each fragment calculated to include one gene or perhaps a few. We "recombined" the fragments of *Aplysia* DNA with the DNA of a bacterial virus, phage lambda, and packaged the recombinant DNA in the phage's protein coat. The hybrid phages served as vectors for introducing the small DNA into bacteria and cloning it. When the phages are added to a bacterial culture, each phage infects a single bacterial cell and multiplies, killing the cell; the phage progeny keep multiplying, killing adjacent cells and creating a plaque, or

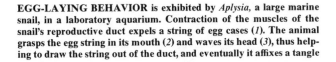

EGG-LAYING BEHAVIOR is exhibited by *Aplysia*, a large marine snail, in a laboratory aquarium. Contraction of the muscles of the snail's reproductive duct expels a string of egg cases (1). The animal grasps the egg string in its mouth (2) and waves its head (3), thus helping to draw the string out of the duct, and eventually it affixes a tangle **of string to a solid substrate (4). In the instance depicted in these drawings the behavior was elicited in an unmated snail by injecting the animal with an extract of *Aplysia* bag cells: neurons in which a "polyprotein" is synthesized from which an egg-laying hormone (ELH) and other peptides associated with egg-laying behavior are cleaved.**

hole, in the culture. Each plaque is occupied by millions of phages descended from an individual phage and therefore contains a clone of millions of copies of a single *Aplysia* DNA fragment.

In selecting a probe with which to scan the clones and find the ELH gene we relied on the fact that ELH is synthesized in abundance in the bag cells, which must therefore contain a large amount of ELH messenger RNA. This calls for some explanation. DNA is a double-strand molecule each of whose strands is a chain of four different nu-

cleotide subunits. Genetic information is encoded in the sequence of nucleotides defining a gene. Groups of three nucleotides are codons, or code words, each of which specifies a different one of the 20 amino acids that are the subunits of proteins. The DNA is not translated direct-

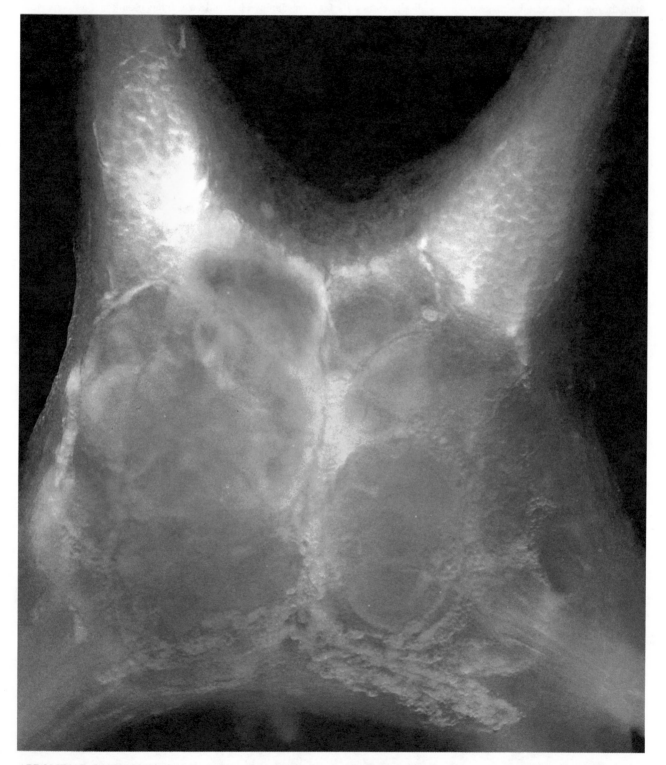

ABDOMINAL GANGLION of *Aplysia* is enlarged about 40 diameters in this photomicrograph. The two clusters of bag cells, where ELH is synthesized, are visible lying athwart the large nerve bundles at the top left and top right. Within the main body of the ganglion one can discern a number of very large neurons, or nerve cells, many of which have been individually identified and found to be invariant in all members of the species. ELH and its companion peptides have been shown to have specific effects on the firing of certain neurons.

ly into protein, however. The coding strand of DNA is first transcribed into a complementary strand of the similar nucleic acid messenger RNA whose sequence reflects that of the gene, and the RNA is translated into protein according to the genetic code.

Since ELH messenger RNA is predominant in the bag cells but not in non-neural tissues, we isolated messenger RNA from bag cells and from non-neural cells. We "reverse-transcribed" it into complementary DNA and labeled the DNA with a radioactive isotope. The labeled DNA was then exposed to the library of recombinant clones under conditions such that the ELH complementary DNA would "hybridize" with, or bind to, stretches of cloned genes encoding ELH; the non-neural complementary DNA would not do so. Clones with which the radioactive ELH probe hybridized but with which the non-neural probe did not hybridize were revealed by autoradiography.

In this way we identified a number of clones containing ELH genes. (As we shall explain, there turned out to be three somewhat different genes, but they are so similar that the bag-cell probe hybridized with all of them.) Having identified clones containing the ELH gene, we grew some of them up into large cultures, isolated their recombinant DNA and exposed it to ELH messenger RNA. In electron micrographs we could detect the exact sites on the long recombinant-DNA molecule at which the RNA hybridized with the complementary DNA and so could close in on the precise stretch of DNA constituting the ELH gene.

Having isolated the ELH gene, we sequenced it, that is, we determined the order of its nucleotides. Given the nucleotide sequence, we could "reverse-translate" it according to the genetic code and thereby deduce the amino acid sequence of the protein chain encoded by the gene. The gene's protein product, it turned out, must have 271 amino acid subunits. Yet we knew that the ELH hormone itself has only 36 amino acids, and indeed we could see the small ELH sequence within the longer protein sequence. The ELH peptide is flanked on both sides by the same pair of amino acids: lysine and arginine. The lysine-arginine pair is known to be a cleavage signal. It serves as a site at which specific enzymes cut a large precursor protein chain to make an active smaller protein or peptide. The most obvious explanation, then, was that the 271-amino-acid chain is a precursor molecule from which ELH is cleaved.

It was surprising, however, that the ELH peptide accounted for such a small part of the ELH gene's protein product. Was the rest of the precursor

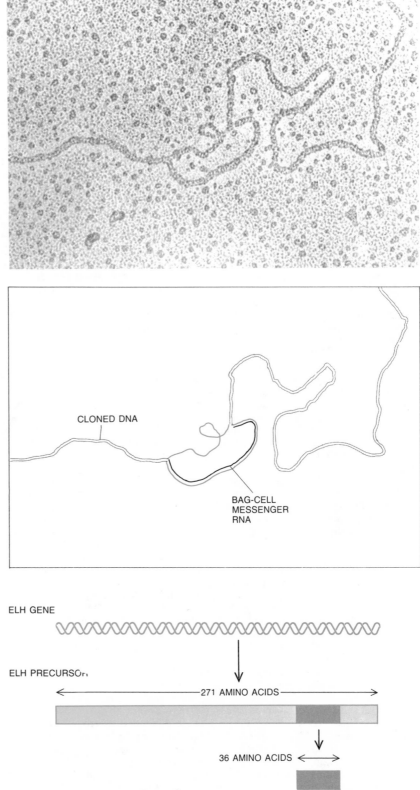

ELH GENE is demarcated in an electron micrograph (*top*). Fragments of snail DNA were recombined with the DNA of a bacterial virus and cloned. A clone containing the ELH gene was identified, expanded and then exposed to ELH messenger RNA under conditions promoting the formation of RNA-DNA hybrids. As is shown in the map, the RNA (*black*) hybridized with the complementary strand of the ELH gene (*color*), disrupting the cloned-DNA duplex. When the ELH gene was sequenced, it was seen to encode a precursor protein chain 271 amino acids long. The 36-amino-acid ELH peptide was identified within the precursor protein.

chain simply cast off unused or might it too contain biologically active peptides? We examined the sequence for additional cleavage signals. It turned out that there are 10 such signals in all; if each site were recognized and cut, 11 discrete peptides would be generated from the single ELH precursor protein. This raised the exciting possibility that such peptides might be other elements controlling egg-laying behavior.

Earl M. Mayeri and his colleagues at the University of California at San Francisco School of Medicine had been investigating the physiological properties of neuropeptides released by the bag cells, and so we joined them in an effort to see whether any of the peptides predicted from the ELH-gene sequence were present in bag-cell extracts and, if they were, whether they would prove to be active in the nervous system. Mayeri and Barry S. Rothman identified and sequenced three other small peptides in the bag-cell clusters: alpha bag-cell factor, beta bag-cell factor and acidic peptide. We found that each of them, bounded by cleavage sites, is encoded along with ELH in the gene we had isolated. Electrophysiological experiments showed that three of the four peptides (ELH and alpha and beta bag-cell factors) interact with specific identifiable neurons in the abdominal ganglion, where each acts as a neurotransmitter: a molecule that mediates the transfer of electrical activity from one neuron to another.

ELH acts locally as an excitatory transmitter, augmenting the firing of the abdominal-ganglion neuron designated $R15$. In addition to acting as a neurotransmitter the ELH peptide diffuses into the circulatory system and excites the smooth-muscle cells of the reproductive duct, causing them to contract and expel the egg string. In other words, ELH acts not only as a neurotransmitter but also as a hormone. (The peptide's effect on the duct muscle was noted before its nervous-system activity; hence the name egg-laying hormone.) The beta bag-cell factor is also an excitatory transmitter. It causes the firing of two symmetrical neurons, $L1$ and $R1$, whose functions are not known. The third neuropeptide, alpha bag-cell factor, is an inhibitory transmitter. It inhibits the firing of a cluster of four neurons, $L2$, $L3$, $L4$ and $L6$. In addition it appears to have a feedback-amplification effect: it is capable of exciting the bag cells from which it has been released.

The association of these three individual peptides, encoded by a single gene, with the activity of different sets of neurons (and muscle cells) suggests that an interesting mechanism may be responsible for generating the complex array of behaviors associated with egg laying. A single gene appears to specify a "polyprotein": a protein chain that is cut into a number of small, biologically active peptides. Perhaps all the components of the egg-laying behavior are mediated by peptides encoded by this one gene. If that is the case, there would be an all-or-nothing effect: no one component of the behavior would be displayed in the absence of the others. Moreover, the synthesis of the single polyprotein would rigidly coordinate the timing of a battery of related behaviors. A single gene encoding multiple neuroactive peptides might thus dictate a complex array of innate behaviors: a fixed-action pattern.

If indeed peptides derived from a single polyprotein mediate the egg-laying behaviors, what controls the release of the peptides to initiate the behavioral array at an appropriate time in the animal's life? The release of ELH and the other putative egg-laying peptides takes place after prolonged electrical excitation of the bag cells. Strumwasser's group has isolated and sequenced two peptides, designated A and B, that are synthesized in the atrial gland, an organ in the snail's reproductive system. Injection of the atrial-gland peptides into an animal results in the excitation of the bag-cell clusters, prompting the release of ELH and its companion bag-cell peptides. Whether the A and B peptides actually initiate the egg-laying process in nature has not yet been demonstrated, but in the laboratory they do appear to be factors controlling the release of the bag-cell peptides.

When we examined the amino acid sequence of the A and B peptides, we noticed that short blocks of amino acids in our ELH precursor were homologous to (the same as or almost the same as) stretches of the atrial-gland peptides. The reader will recall that in our cloning experiments the probe made from ELH messenger RNA had hybridized to three somewhat different DNA's, only one of which encoded ELH and its companion bag-cell peptides. It was possible that the three DNA's were members of a multigene family, with one of them encoding ELH and the other two encoding the A and B peptides.

In a series of hybridization experiments we were able to show that messenger RNA derived from the two genes related to the ELH gene is indeed synthesized in the atrial gland, where Strumwasser had found the two peptides. We determined the nucleotide sequence of the two genes and, by reverse translation, the amino acid sequence each of them encodes. We saw immediately that although the two genes are closely related to the ELH gene, they do not encode the ELH peptide that is active in the bag cells. Instead they encode the A and B peptides of the atrial gland. The three genes are clearly members of a small multigene family with a common evolutionary origin, but they have diverged to generate different—yet functionally related—sets of peptides.

The similarities and the differences are revealed when the three nucleotide and amino acid sequences are compared in detail [*see illustrations on page 8*]. Each gene encodes a precursor protein in which lysine-arginine sites (or sometimes a single arginine or two adjacent arginines) delimit the blocks of amino acids that are cleaved to become active peptides. All three precursors begin with a characteristic signal sequence of about 25 amino acids that governs the processing of the protein chain. The newly translated chain enters the lumen of a membrane system called the rough endoplasmic reticulum, where it begins to be modified: the signal sequence is cut off and sugar and phosphate molecules are added to the protein, which proceeds to the organelle called the Golgi apparatus. There the precursor is cleaved and the component peptides are enclosed in small vesicles, or sacs. In response to appropriate stimuli the vesicles fuse with the outer membrane of the secreting cell and release their contents to interact with nearby cells, to diffuse through the ganglion or to enter the circulation.

The differences between the precursors incorporating the A and B peptides and the precursor incorporating ELH begin after the signal sequence. Let us first describe the A and B precursors. A single-arginine site signals the beginning of either the A or the B peptide, both of which consist of 34 amino acids. At the end of these peptides there is a glycine-lysine-arginine sequence, which serves as a signal not only for cleavage but also for transamidation: the addition of an amino group (NH_2) at the end of the peptide, replacing the usual hydroxyl group (OH). Transamidation "blocks" the end of the peptide, perhaps making it more resistant to degradation. There follows, in both the A and the B precursor, a stretch of 47 amino acids unrelated to any known peptide. Then comes another lysine-arginine cleavage site, followed by what looks like the beginning of the ELH peptide.

ELH is not synthesized in the atrial gland, however. Examination of the nucleotide sequence of the atrial-gland genes shows why. In the case of the A-peptide gene, the first 22 amino acids of ELH are encoded correctly. Then a single-nucleotide difference in the codon for the amino acid at position 23 generates an arginine-arginine-arginine sequence and thus establishes a potential cleavage site that could break up what would otherwise become the ELH peptide. A different kind of single-nucleotide change is seen in the B-peptide gene.

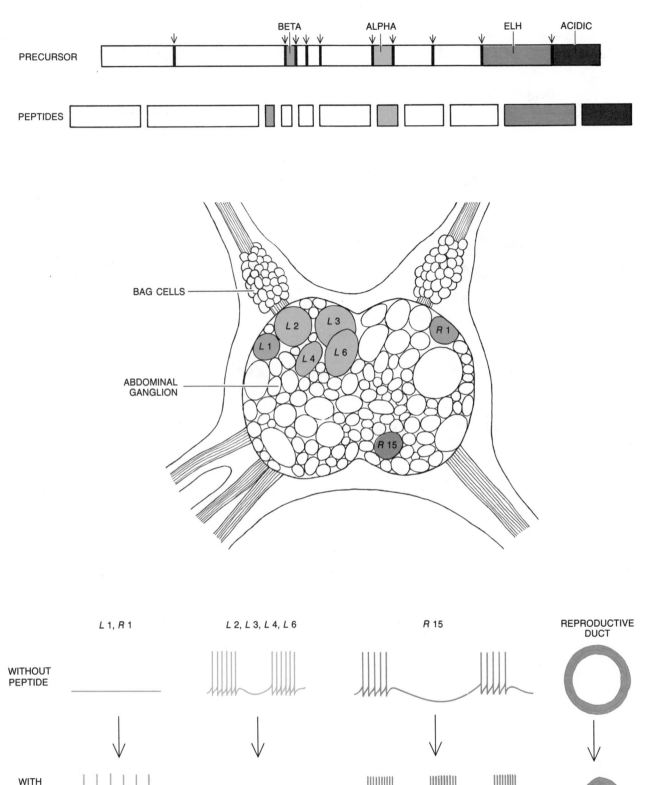

ELH PRECURSOR was found to be a polyprotein containing a number of active peptides. The precursor (*top*) is studded with 10 sites (*arrows*) at which a protein chain is cleaved by enzymes called endopeptidases. Cleavage at all the sites would release 11 peptides (*second from top*). Four of those peptides are known to be released by the bag cells: the beta and alpha bag-cell factors, ELH and acidic peptide.

Three of them (*colored peptides*) have been shown to act as neurotransmitters, altering the activity of specific abdominal-ganglion neurons (*colored cells*) in specific ways (*bottom*). The beta factor excites cells *L*1 and *R*1. The alpha factor inhibits cells *L*2, *L*3, *L*4 and *L*6. ELH augments the firing of cell *R*15. ELH also enters the circulation and acts as a hormone, causing contraction of the reproductive duct.

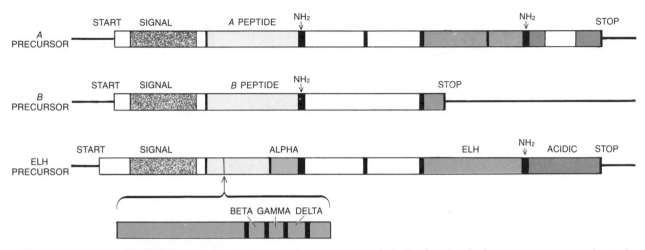

THREE PRECURSOR PROTEINS encoded by the three members of the ELH multigene family are compared. These precursors are cleaved at specific sites (*black bars*) to give rise to active peptides, some of which undergo amidation (*NH₂*). The three precursors have in common a signal sequence (*left*) and several regions of homology, or close similarity, that give rise in one precursor or another to the *A* or the *B* peptide or to ELH and acidic peptide. Single-nucleotide differences alter the *A* and *B* genes so that active ELH is not synthesized. An 80-amino-acid insertion in the ELH gene interrupts the *A*- and *B*-peptide sequences and gives rise to several different peptides.

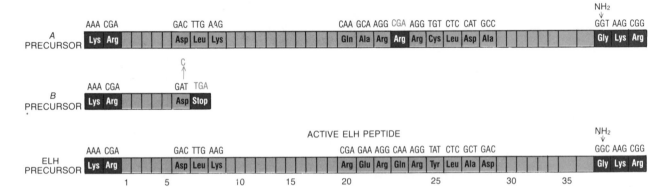

ELH HOMOLOGY REGION is very similar in all three genes, that is, the sequence in which the four nucleotides (*A, G, C* and *T*) are arrayed is almost the same, so that the three-nucleotide codons specify the same amino acids at most positions. (Only some of the more significant codons and amino acids are shown here.) Whereas the ELH gene gives rise to an active ELH peptide, the other two genes do not. In the *A*-precursor gene the substitution of a *G* for an *A* nucleotide at position 23 gives rise to an arginine instead of a glutamine, generating a three-arginine sequence and thus a potential cleavage site. In the *B*-precursor gene a *C* nucleotide is deleted from the sixth codon, changing the "reading frame" so that a *TGA* is introduced at position 7. *TGA* serves as a "stop" codon, which terminates translation.

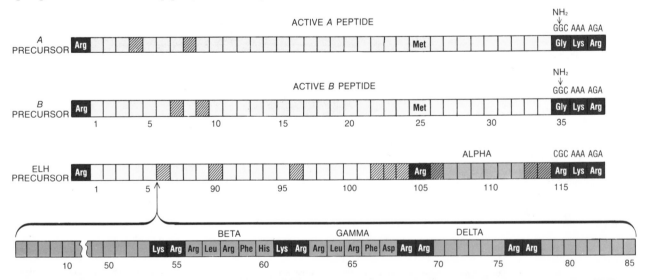

EIGHTY-AMINO-ACID INSERTION disrupts the *A*- and *B*-peptide homology region in the ELH gene. It is cleaved at lysine-arginine sites to make three peptides: the beta, gamma and delta bag-cell factors. (Note the near-identity of beta and gamma, which presumably arose by duplication of a short DNA sequence.) At position 105 in the homology region the substitution of an arginine in the ELH precursor for the methionine of the *A* and *B* precursors creates a cleavage site, giving rise to another active peptide: alpha bag-cell factor. Apart from the insertion the three precursors are very similar. The hatching marks sites where one of them has a different amino acid.

Here one nucleotide in the sixth codon is deleted. The "reading frame" of nucleotide triplets is thereby changed so that a "stop" codon is generated; translation is terminated after only a six-amino-acid stub of ELH has been synthesized.

Consider now the ELH precursor that is synthesized in the bag cells. The nucleotide sequence of its gene is very similar to that of the gene encoding the *A* and *B* peptides, and yet the gene does not specify those peptides; it specifies ELH and a number of other peptides involved in egg laying. It begins with the same signaling sequence seen in the *A* and *B* precursors. Then come the first five amino acids of the *B* peptide. At this point, however, the ELH precursor diverges dramatically from the *A* and *B* precursors. The ELH gene contains a 240-nucleotide sequence that is not present in the *A* or the *B* gene, encoding 80 amino acids. The insert includes four cleavage signals delimiting three bag-cell factors: beta, gamma and delta. The beta factor, as we have mentioned, is known to have a specific effect on abdominal-ganglion neurons *L*1 and *R*1.

After the insert the nucleotide sequence resumes, without any alteration of the reading frame, to encode the sixth amino acid of the *B* peptide, and then it continues through a sequence that is much like the sequence of the two atrial-gland genes. One divergence in this region is particularly significant. A cleavage site is introduced, generating the nine-amino-acid alpha bag-cell factor that, as described above, inhibits the firing of four neurons in the abdominal ganglion.

There follows a stretch of incomplete homology with the *A* and *B* genes, after which a cleavage site signals the beginning of the 36-amino-acid ELH peptide. The end of this peptide, as in the case of *A* and *B,* is followed by a signal for cleavage plus transamidation. Between the end of the ELH peptide and the stop codon that puts a halt to translation there remain 27 amino acids: those of acidic peptide, which is also released from the bag cells along with ELH but whose target is not yet known.

Here, then, are three genes that have in common three regions of homology: the *A* or *B* region, the ELH region and the acidic-peptide region. Moreover, within each of these three peptide regions there are near-identities of sequence at fixed positions. The implication is that all three peptides had their origin in a small ancestral peptide whose gene triplicated to generate a larger protein composed of at least three peptides. The gene encoding that larger protein apparently triplicated in turn, giving rise to three independent genes that diverged as they became specialized to satisfy different functional requirements. Then

there may have been minor duplications of some regions of one or another gene, as is suggested by the fact that the adjacent beta and gamma peptides are almost identical. Such events presumably allowed for the evolution of variants without significant alteration of the original gene. The various versions of the gene may have been transposed to different sites in the genome, perhaps on different chromosomes.

The availability of cloned ELH genes enabled us, working with McAllister and Kandel, to ask whether the genes are expressed (transcribed into messenger RNA that is then translated into protein) not only in the bag cells and the atrial gland of adult snails but also in other parts of *Aplysia*'s nervous system, and to trace the development of the neurons in which the gene is expressed. We do this by means of two techniques: in situ hybridization and immunocytochemistry. The former depends on the fact that wherever a particular gene is expressed one finds the messenger RNA transcribed from it. We cut a thin section of tissue, mount it on a microscope slide and treat it so that the RNA in its cells is accessible to molecular hybridization. When ELH genes labeled with a radioactive isotope are applied to the slide, they hybridize with the complementary messenger RNA, whose location is revealed by autoradiography. Immunocytochemistry, on the other hand, reveals the presence of the peptide. An antibody to the ELH peptide is applied to a slide and binds to the peptide if it is present. A second antibody, chosen to bind to the first one and linked to a fluorescent dye, is applied, and inspection under ultraviolet illumination reveals the cellular location of the peptide.

We have done hybridization and immunofluorescence studies of the snail's entire central nervous system. The two techniques detect individual neurons expressing an ELH gene even in ganglia with several thousand nonexpressing cells. The bag cells and the atrial gland are clearly the major sites of expression of the ELH genes, but the experiments show that ELH messenger RNA and the peptide itself are also synthesized in an extensive network of nerve cells, not only in the abdominal ganglion but also in three other ganglia. In similar experiments Strumwasser and Arlene Y. Chiu have also identified neurons outside the bag-cell clusters that synthesize ELH. It appears that ELH may have an extensive role as a neurotransmitter throughout the *Aplysia* nervous system.

The presence of scattered ELH-producing cells in four of the ganglia raises the question of how the cells arise during development. By in situ hybridization in developing animals we have learned when the ELH genes are first

expressed and where the neurons expressing ELH originate. In all animals the nervous system develops as a specialization of the ectoderm, the embryonic body surface. The ELH genes are active very early in the larval stages of the snail's development, in a zone of primitive cells that line the body wall and are destined to become neurons. Later in development these cells leave the body wall and migrate, by crawling along strands of connective tissue, to their ultimate locations in the adult nervous system. It may be that a single primitive neuron divides to generate a cluster of ELH-producing cells, which migrate not only to the bag cells but also to scattered sites throughout the nervous system.

In *Aplysia* at least three genes encode a number of neuroactive peptides that have roles in the circuits governing a complex but stereotyped behavioral repertoire. In the mammalian brain too a number of peptides have been identified that seem to mediate specific behaviors [see "Neuropeptides," by Floyd E. Bloom; SCIENTIFIC AMERICAN Offprint 1502]. What properties of neuropeptides, and in particular what characteristics of their mode of synthesis, suit them to a role as mediators of behavior?

The behavioral potential of an organism is at least in part dictated by a rigid network of connecting nerve cells. Much of the communication between neurons is local and is mediated by a neurotransmitter such as acetylcholine or norepinephrine, which is released from a neuron, crosses the narrow synaptic space and makes point-to-point contact with another neuron. Neuroactive peptides too can function locally as neurotransmitters. In addition, however, they can be secreted into the circulation to serve as neurohormones that act on several distant targets and thus generate several discrete activities. Peptides thereby make available an additional communication network that supplements the hard-wired network of interconnected nerve cells.

The diversity of a neuropeptide's target sites can enable the peptide to coordinate physiological events with particular behaviors. For example, the injection of the peptide angiotensin II elicits spontaneous drinking in vertebrates by acting on neurons of the hypothalamus in the brain. The peptide also acts indirectly on the kidney to promote the reabsorption of sodium and water into the bloodstream instead of their excretion. Both of these two quite different effects serve to rehydrate the animal. ELH provides another example. In *Aplysia* it acts locally to excite specific neurons in the ganglia that may cause such behaviors as head waving and such physiological changes as an increase in heart rate. At the same time it acts at a distance, causing the reproductive duct to contract and expel eggs.

A number of egg-laying peptides of *Aplysia* are cleaved from a single polyprotein precursor. Some other neuropeptides are known to be synthesized in the same way. The polyprotein pathway would seem to have several advantages. For one thing, it provides a simple mechanism for the control of both peptide synthesis and release. Different peptides encoded by a single gene can be synthesized simultaneously under the control of a single regulatory agent. Moreover, the various small peptides of a polyprotein may, after cleavage from the precursor, be packaged in the same vesicle, to be released from the synthesizing cell simultaneously by a single stimulus.

The generation of multiple peptides from one precursor also provides a solution to a number problem: the number of genes in an animal genome is simply not large enough to specify the diversity of behaviors exhibited by a species. The information potential of a single gene can be increased if the gene's protein product is cleaved differently in different cells or in response to different stimuli. An example of alternate protein processing has been noted by Edward Herbert of the University of Oregon and James L. Roberts of the Columbia College of Physicians and Surgeons and also by Richard E. Mains and Betty A. Eipper of the Johns Hopkins University School of Medicine. They found that a single precursor is processed to make adrenocorticotrophic hormone (ACTH) in the anterior lobe of the pituitary gland and to make an endorphinlike peptide in the posterior lobe. We have observed one pattern of cleavage of the ELH precursor in a single cell. In principle, given the cleavage sites we have noted, the protein could be cut in different ways to generate more than 2,000 different combinations of peptides, and each combination could activate a different pattern of behavior. The potential for diversity is greater still because the ELH genes constitute a family of genes expressed in different tissues.

The polyprotein pathway offers temporal flexibility by allowing the various peptides to remain active for different lengths of time. The stability of a peptide can be influenced by certain postsynthetic modifications such as transamidation and even by size (because small peptides cannot easily fold into compact structures and are therefore likely to be degraded sooner than large ones). Both the pattern of amidation and the size of the various egg-laying peptides are consistent with their functional requirements. The peptides (ELH and the *A* and *B* peptides) that can act as hormones are consistently amidated and are longer, and so their active life is presumably extended. The peptides (such as the alpha and beta bag-cell factors) that seem to act only locally as neurotransmitters are shorter and are not amidated. They are presumably degraded faster, like most other neurotransmitters.

The organization of genes encoding polyproteins provides a striking degree of evolutionary flexibility. The interspersal of sequences encoding a set of active peptides in a gene that also encodes nonfunctional protein leaves room for the evolution of new active peptides without alteration of the original set. The ELH gene, for example, has a 240-nucleotide insertion that is not present in the homologous genes expressed in the atrial gland. The insertion encodes three peptides (two of which appear to reflect a small internal duplication), thus expanding the polyprotein's array of coordinated peptides without affecting the synthesis of active ELH.

Finally, the same peptide may be incorporated in several different precursors encoded by different genes. Consider head waving in *Aplysia*. A characteristic waving of the snail's head takes place during feeding as well as during egg laying. The same peptide or peptides could elicit the same behavioral component (head waving) in two very different contexts. To this end the head-waving peptide (or peptides) may be encoded in some other gene—one implicated in feeding behavior—as well as in the ELH gene. In this way complex behaviors could be assembled by the combination of simple units of behavior, each unit mediated by one peptide or a small number of peptides.

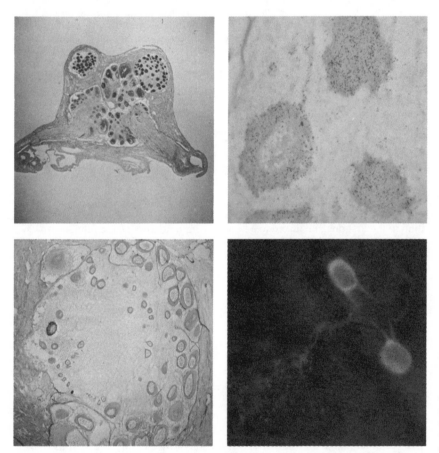

SITES OF ELH SYNTHESIS are revealed by in situ hybridization with the peptide's messenger RNA (*top left and top right, bottom left*) and the binding of antibodies to the peptide itself (*bottom right*). A section of the abdominal ganglion was exposed to an ELH-gene probe labeled with a radioactive isotope (*top left*). The probe hybridized with ELH messenger RNA, whose location was revealed by autoradiography. The black dots show hybridization in the bag cells (rounded clusters at the upper margin of the ganglion) and also in a single neuron near the middle of the ganglion. The magnification is about 17 diameters. Magnification to about 400 diameters shows ELH messenger RNA in the cytoplasm of bag cells and their processes (*top right*). ELH messenger RNA is also detected in a single cell of a different group of neurons, the pleural ganglion (*bottom left*). In an immunofluorescence study (*bottom right*) an antibody to ELH was applied to a section taken from the abdominal ganglion. A second antibody, directed against the anti-ELH one and labeled with a fluorescent dye, was applied. The bright orange fluorescence shows ELH in two abdominal-ganglion cells and their processes.

Lithium and Mania

by Daniel C. Tosteson
April 1981

*How is it that salts of lithium have a beneficial effect on
people in a pathologically manic state? Clues to the
answer may be found in the ways the lithium ion moves
through the membranes of cells*

The word mania applies loosely to many kinds of excitement, but its clinical meaning is more specific. There the term describes an illness in which the patient is mentally and physically hyperactive and has an elevated mood and disorganized behavior. Often the illness has an obverse side, depression, and the patient's mood swings periodically from one extreme to the other, a syndrome usually described as a bipolar disorder. People suffering from mania or bipolar disorders often improve impressively when they are treated with salts of lithium.

These salts, like most others, dissociate into positively and negatively charged ions when they are dissolved in water. The distribution and many of the actions of lithium ions in the human body involve interactions with the membranes that envelop cells and intracellular organelles. Indeed, the transmission of impulses between and along nerve cells in the brain depends on the regulated movement of ions across the membranes of those cells. It has recently become clear that some patients with mania have a defect in a pathway for the transport of lithium across the membranes of the red cells of the blood. A plausible conjecture is that the condition also affects brain cells. I shall set forth in this article what is now known about the relations among lithium, membranes and mania. I shall also venture some speculations about the significance of this knowledge for deepening human understanding of the physicochemical basis of the human mind.

The story begins in 1949, when a brief report by J. F. J. Cade appeared in *The Medical Journal of Australia* telling of a marked improvement of patients with acute and chronic mania who had been treated with lithium salts. At that time Cade, a physician and psychiatrist, was working alone at a small hospital for chronic mental patients and had not had any previous experience in research. His interest was in the conditions that give rise to such affective disorders as mania and depression. He reasoned that the manic state might be initiated and

sustained by a substance that could be found in the body fluids of people who had the disorder.

Cade began a search for such a hypothetical compound by injecting urine from patients with mania and depression and from normal people into the peritoneal cavity of guinea pigs. Because uric acid and its salts can be present in high concentration in urine he chose to investigate the effect of injecting uric acid into the animals. And because uric acid and many of its salts are not readily soluble in water he worked with lithium urate, which is more soluble. In order to control for any effects of lithium as opposed to urate he injected a solution of lithium carbonate into other guinea pigs. To his surprise he found that after a latent period of about two hours the animals became extremely lethargic and unresponsive, remaining in that condition for an hour or two before again becoming timid and active as they are normally. "Those who have experimented with guinea pigs," Cade later wrote, "know to what extent a startle reaction is part of their makeup. It was thus even more startling to the experimenter to find that after an injection of a solution of lithium carbonate they could be turned on their backs and that, instead of their usual frantic righting behavior, they merely lay there and gazed placidly back at him."

Cade speculated that the calming effect of lithium on guinea pigs might be repeated in human beings. He therefore began a study of the effect of administering lithium salts orally to a 51-year-old man who had been hospitalized with chronic mania for five years. After five days of treatment the patient showed a definite improvement; within two months he was able to leave the hospital and resume his job while remaining on lithium therapy.

After Cade's initial report the role of lithium therapy in mental disorders was most vigorously explored by Mogens Schou and his associates at the University of Aarhus in Denmark. By the early 1960's lithium was accepted as the

treatment of choice for mania in Australia and most of Europe. Another 10 years passed before lithium was accepted in the U.S. One reason for the long delay was the great caution that many American physicians had about lithium because of disastrous experience with it as a substitute for sodium in the treatment of congestive heart failure in the 1940's. Several patients had died before the toxicity of lithium, which is particularly pronounced in a person deprived of sodium, had become evident. Also contributing to the delay were economic factors. Because the chemical simplicity and ready availability of lithium made it impossible for pharmaceutical companies to develop an exclusive position for the marketing of the drug, they had little incentive to make lithium salts and to carry out all the testing required by the Food and Drug Administration before lithium could be introduced into clinical practice in the U.S.

Lithium is the third-lightest element, following hydrogen and helium in the periodic table of the elements. The atom has three protons (positive charges) in its nucleus and three electrons (negative charges) in orbit around the nucleus. One of the electrons is alone in the outer orbital and so is readily lost to interactions with other atoms. When it is lost, the lithium atom has three positive charges in its nucleus and two negative charges in its electron shell and is hence a cation: a positively charged ion that interacts with anions, or negatively charged ions. In the synthesis of heavier chemical elements out of lighter ones in the universe lithium is not on the main pathway; therefore on the earth it is present in small quantities compared with its chemical relatives sodium, potassium, magnesium and calcium. Nevertheless, it is distributed widely in minerals, largely as silicates.

From the point of view of the physical chemist the striking thing about lithium, sodium and potassium is their similarity. Indeed, it was this similarity that led to their classification in Group IA of the periodic table. From the viewpoint of the physiologist, however, the signifi-

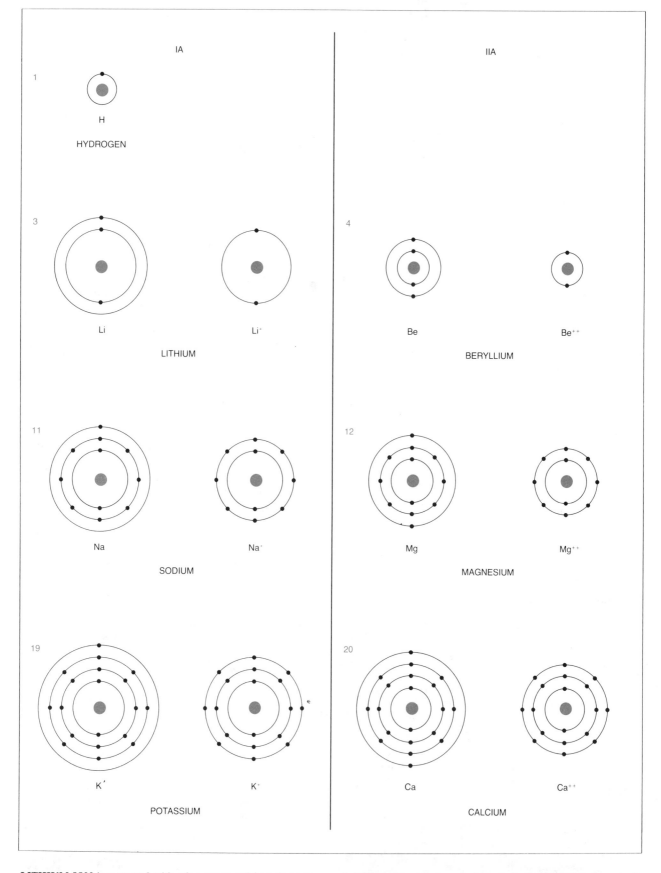

LITHIUM ION is portrayed with other atoms and ions that move through cell membranes. They are presented according to their positions in groups IA and IIA of the periodic table of the elements and also according to their relative sizes. The colored numerals show the number of protons, or positive charges, in the nucleus of each atom; the black dots represent electrons in their orbits around the nucleus.

A lithium atom has two electrons in its first orbital and one in its second: a total of three negative charges balancing the three positive charges of the nucleus. The single electron is readily lost, however, and the atom becomes a cation: a positively charged ion (Li⁺). A number of patients with clinical mania can be restored to normality by treatment with lithium salts, which in body fluids dissociate into ions.

cance of these elements lies in their differences.

The ability of the membranes of the living cell to generate and propagate electrical signals depends on the ability of these membranes to distinguish between sodium and potassium. The two elements must also be recognized and separated by the molecular machine that pumps sodium out of and potassium into nerve and muscle cells and thereby charges the "battery" that drives the electrical signals. The effects of lithium in living organisms, including human beings with affective disorders such as mania, turn on the differences between ions of the element and ions of the other members of groups IA and IIA that are present in cells, particularly sodium, potassium, calcium and magnesium. Biological evolution on the earth apparently has taken advantage of subtle differences among some of the elements that make up the crust of the earth.

The distribution and rates of movement of ions in living organisms are largely regulated by the membranes that separate the inside and the outside of cells and of intracellular organelles such as the mitochondrion. Such a membrane consists of a bilayer of phospholipid and cholesterol molecules in which specialized protein molecules are embedded. The thickness of the membrane, including the proteins, is about 10 nanometers (10 millionths of a millimeter).

The lipid-bilayer part of these biological membranes readily allows the passage of small, uncharged, relatively lipid-soluble molecules such as those of carbon dioxide and molecular oxygen but not the passage of charged, water-soluble ions such as those of sodium, potassium and lithium. Most if not all of the transport of ions across a biological membrane is mediated by specific proteins embedded in the lipid bilayer. Four types of transport deserve mention. First, protein channels may allow the selective movement of various ions down their respective concentration and electric-potential gradients. One example is provided by the time- and voltage-dependent sodium and potassium channels that are responsible for the propagation of the action potential (the nerve impulse) in nerve and muscle fibers. Another is seen in the channels, activated by acetylcholine, that initiate the action potential in muscle fibers.

Second, there are pumps that convert the energy of chemical bonds, such as the energy stored in the pyrophosphate bond of adenosine triphosphate (ATP), into the work necessary to move ions against concentration and electric-potential gradients. Among the examples are the sodium-potassium pump present in the outer membrane of most cells and the calcium pumps present in the membranes of mitochondria and the sarcoplasmic reticulum.

A third kind of ion transport was discovered by Hans H. Ussing of the University of Copenhagen, who called it exchange diffusion. The process allows an exchange but no net movement of ions through the membrane. Such transport was first observed by Ussing and his colleagues as sodium-sodium exchange in frog muscle. To detect an exchange involving a single species of ion (sodium in Ussing's observation) requires a tracer such as a radioactive isotope of the transported ion. If two different kinds of ion (such as sodium and lithium) can participate in a system of this kind, the exchange proceeds until the ratio of concentrations of the two ionic species is identical in both solutions (inside and outside) bathing the membrane. Chloride ions and bicarbonate ions are exchanged across the membranes of human red blood cells by this kind of exchange mechanism.

A fourth type of molecular movement across biological membranes is cotransport, in which two substances move together. Examples are the simultaneous movement of sodium and sugars or sodium and amino acids into the epithelial cells of the intestine as part of the process of absorption and the coupled cotransport of sodium and potassium in the red cells of birds and in certain other cells. Lithium can substitute for sodium in all these pathways. Under the conditions that prevail in the blood of patients being treated with lithium, however, only one pathway—sodium-lithium exchange or countertransport—moves lithium outward across the red-

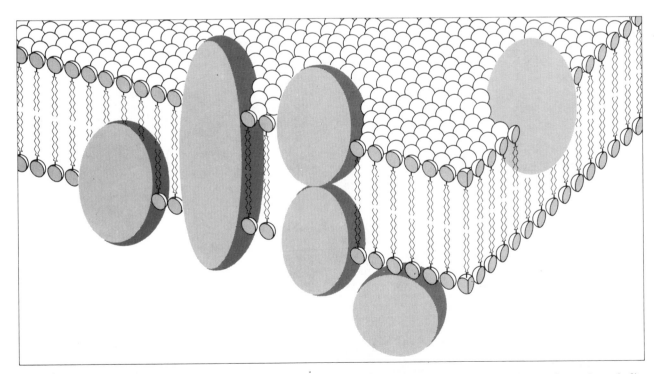

STRUCTURE OF THE CELL MEMBRANE is shown schematically according to current conceptions. The basic structure is a double layer of lipid (fat) molecules with their hydrophilic heads pointing outward and their hydrophobic tails pointing inward. Also associated with the membrane are protein molecules, seen here as larger bodies some of which penetrate the membrane and others of which are embedded in one side or the other. Most of the movement of ions into and out of the membrane is by way of channels in the various proteins.

cell membrane against its concentration gradient.

In addition to regulating the distribution of ions and small molecules, membranes carry out many other important cellular functions. Among them are the uptake and release of large molecules and small volumes of fluid; the recognition of extracellular molecules such as hormones, antigens and antibodies and of other cell surfaces in embryonic development and the repair of organs, and the ordering of the sequence of various biochemical reactions.

Some of the connections among lithium, membranes and mania began to become clearer when Joe Mendels and

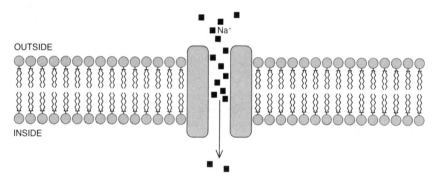

PROTEIN CHANNEL through which ions such as sodium, potassium and lithium can move "downhill" involves a penetrating protein. Downhill movement, which can be either into or out of the cell depending on differences in the ion concentration and the electric potential on the two sides of the membrane, does not require metabolic energy. Here sodium ions move inward.

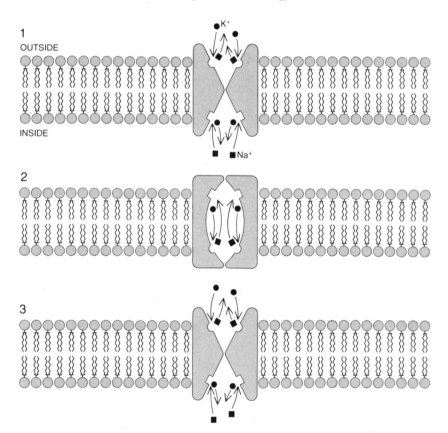

ION PUMP provides another means of moving ions into and out of a cell. It converts the energy of chemical bonds into the work necessary to move ions against their electrochemical-potential gradients. A two-state, four-site speculative mechanism for the sodium-potassium pump begins (1) with a membrane-penetrating protein that is open to the outside and inside of the membrane. On the outside a potassium ion from the blood plasma moves to a binding site vacated by a sodium ion; a reverse exchange takes place on the inside. The configuration of the protein channel changes to a closed state (2), the selectivity of the two sites changes and potassium on the outward-facing site exchanges with sodium on the inward-facing site. The first configuration and selectivity of sites are restored (3), and the cycle is completed. The changes of shape and selectivity are driven by the hydrolysis of adenosine triphosphate (ATP). At no time is there an open pathway that allows ions to move the entire distance through the membrane.

Alan Frazer of the University of Pennsylvania School of Medicine observed that the lithium ratio (the concentration of lithium inside red cells divided by the concentration in the blood plasma outside the cells) showed a considerable variation from person to person among patients who were being given lithium therapeutically. Mendels and Frazer also found that the mean value of the ratio in a group of patients with affective disorders responsive to lithium was significantly higher than it was in other groups of patients and in normal people who served as controls.

Mendels and Frazer also reported that the lithium ratio is higher in sheep red cells with a high concentration of sodium than it is in sheep red cells with a low concentration of sodium. This finding led me to suspect that a countertransport system involving sodium and lithium could be found in the membranes of human red cells. Experiments that Mark Haas, James M. Schooler and I did at the Duke University School of Medicine in 1975 supported the hypothesis. Somewhat later, in collaboration there with G. N. Pandey and John M. Davis and their colleagues, with Balasz Sakardi and R. B. Gunn at the University of Chicago and with J. Funder and J. O. Wieth of the University of Copenhagen, we investigated the transport processes that determine the distribution of lithium between human red cells and blood plasma. Simultaneously and independently Jochen Duhm and his colleagues on the medical faculty of the University of Munich made similar observations that were later confirmed and extended by Barbara E. Ehrlich and Jared M. Diamond of the University of California at Los Angeles School of Medicine. On the basis of this now quite substantial body of evidence the following picture of lithium metabolism in human red cells emerges.

The concentration of lithium in the blood plasma of patients receiving treatment for affective disorders is about one millimole per liter (one millimolar). The steady-state concentration of lithium in the red cells of such patients is the result of a dynamic balance between inward leakage down the lithium concentration gradient and outward countertransport up the gradient. The uphill movement is driven not by the release of energy stored in chemical bonds (as in ATP) but by a dissipation of the internally directed concentration gradient of sodium. The sodium that enters the cell through the lithium-sodium exchange system is extruded by the sodium-potassium pump, which is driven by ATP. The countertransport system therefore has the effect of coupling lithium to the sodium-potassium pump. The steady-state lithium ratio approaches the sodium ratio of .08 when the lithium-sodium countertransport system is very active and there is little or no leakage. When the leakage pathway is large and the countertransport pathway is small, the lithium ratio approaches 1.2—the value for passive singly charged cations when the electric potential inside the cells is 10 millivolts negative with respect to the outside plasma, as is the case for human red cells. In most people the lithium ratio is between .2 and .6.

The countertransport system carries out a 1 : 1 exchange of sodium for sodium or of lithium for sodium. Depending on the sodium gradient, the system can move lithium uphill either into or out of the red cells. Since in the body the concentration of sodium inside cells is almost always much lower than the concentration outside them, the countertransport system gives rise to a net outward movement of lithium whenever the lithium ratio is higher than the sodium ratio. The affinity of the system for lithium is about 20 times greater than it is for sodium, as is shown by the fact that the intracellular concentration needed for a half-maximal stimulation of countertransport is .5 millimolar for lithium and 10 millimolar for sodium.

The countertransport system will not accept potassium, ammonium, choline, calcium, magnesium or any other cations so far tested. It is not inhibited by ouabain, which specifically blocks the ATP-driven sodium-potassium pump. It is inhibited by phloretin, a well-known inhibitor of transport systems for sugar and anions. The system does not require ATP or glycolysis (the glucose metabolism that supplies red blood cells with energy).

The maximum transport rate of the lithium-sodium countertransport system varies considerably from individual to individual. The range is from zero to .7 millimole per liter of red cells per hour. Pandey and his colleagues and Duhm and his have shown that these variations correlate with and account for the variation between individuals in the red-cell lithium ratio, measured both in patients receiving lithium and in vitro (with red cells incubated for 24 hours in a plasmalike medium containing 1.5 millimolar of lithium). Considerable evidence from studies of twins and other family members, involving both patients with affective disorders and normal control subjects, suggests that the variations among individuals in the maximum rate of lithium-sodium countertransport are (at least in part) genetically determined.

The downhill leakage of lithium in human red cells can take either of two pathways. One pathway is not inhibited by any compounds yet tested and does not appear to involve other ions or uncharged molecules. The other is an interesting pathway first discovered by Funder and Wieth. Lithium ions (and to a lesser extent sodium ions) can react

with carbonate, a doubly charged anion, to form the singly charged ions lithium carbonate and sodium carbonate. These anions can then be transported on the very rapid singly-charged-anion exchange system that normally carries bicarbonate and chloride ions across the membrane of the red cell. Although the concentrations of lithium carbonate and sodium carbonate and therefore their rates of transport are always quite low compared with other anions that travel on the anion-exchange system, the rate of movement of lithium by this pathway is significant compared with its rate of movement by other pathways. In patients with normal concentrations of chloride and bicarbonate in the plasma about half of the inward leakage of lithium into red cells is through the carbonate pathway.

The movement of lithium across the membrane of the human red cell in a person who is receiving lithium therapy is primarily by way of the leakage pathways and the lithium-sodium countertransport system. Under certain experimental conditions, however, there are other ways for lithium ions to traverse the membrane of the red cell. For example, lithium competes for both the sodium and the potassium sites in the sodium-potassium pump. When sodium and potassium are not present outside the cell, lithium substitutes for potassium and is actively transported into human red cells; when sodium and potassium are absent inside the cell, lithium substitutes for sodium and is actively transported outward. Since the pump has more affinity for sodium and potassium than it does for lithium, not much lithium goes by this pathway in the body because sodium and potassium are always present on both sides of the membrane (and at much higher concentrations than lithium). We have recently shown that lithium can substitute for sodium in the cotransport of sodium and potassium in human red cells. Still oth-

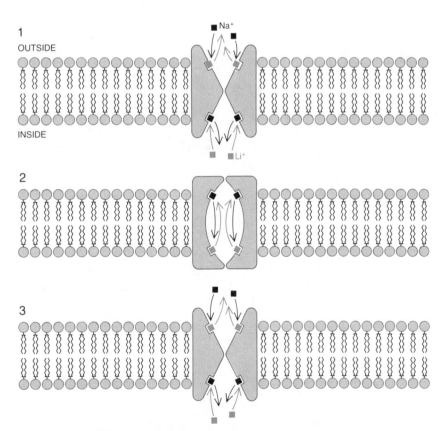

ION-EXCHANGE OR COUNTERTRANSPORT SYSTEM that is thought to contribute to the movement of lithium ions out of red blood cells in therapy for mania is depicted. The system normally carries out a 1:1 exchange of sodium for sodium, but it can also exchange lithium for sodium. The hypothetical two-state, four-site model shown here resembles the pumping mechanism depicted in the lower illustration on the preceding page. In the open state (1) the two sites can exchange sodium or lithium ions present in the solutions bathing the inner and outer surfaces of the membrane. In the closed state (2) the two sites can exchange ions with each other but not with the two solutions. A restoration of the initial state completes the cycle (3). The selectivity of the sites is the same in the two states. In the body the concentration of sodium is lower inside cells than it is outside; the countertransport system produces a net outward movement of lithium when the ratio of cell-to-plasma concentration is greater for lithium than it is for sodium. The countertransport system appears to be defective in certain patients with mania, resulting in a higher cell-to-plasma ratio of lithium than is normal.

er pathways may be important for the transport of lithium in other types of cell. Halvor N. Christensen and his colleagues at the University of Michigan Medical School have recently shown that lithium moves together with certain amino acids in some tumor cells of mice.

What, then, are the connections among lithium, membranes and mania? First, the occurrence of extremely low maximum rates of lithium-sodium countertransport in red cells seems to be greater among people suffering from mania or bipolar disorders than it is among normal people or people with other types of mental disorder. Studies of the families of people with and without affective disorders by Elizabeth Dorus and her colleagues at the University of Chicago Medical School suggest that reduced lithium-sodium countertransport appears more often in the first-degree relatives (parent or child) of patients with mania or bipolar disorders than it does in other people. On the basis of this finding it is possible to entertain the hypothesis that a reduced lithium-sodium countertransport is an inherited disorder and that the genes responsible for the expression are also involved in a predisposition to mania and bipolar disorders. Since most of the people suffering from those affective disorders do not have a reduced lithium-sodium countertransport, it is clear that this hypothesis can apply only to a subgroup of manic and bipolar patients. This is not surprising, because it is well established that an inherited predisposition to affective disorders involves many genes. More extensive studies are needed to elucidate the relation between the inheritability of the lithium-sodium countertransport system and a predisposition to mania and bipolar disease.

If the hypothesis that a reduced lithium-sodium countertransport in red cells and the incidence of mania are positively correlated is correct, what could be the connection between these apparently separate phenomena? It is unlikely that events in the membranes of red cells determine mood. One possibility is that the genes coding for the lithium-sodium countertransport system also express a similar system in other cells, including cells in the brain.

The ionic selectivity of the sodium-exchange pathway in such brain cells might be slightly different, allowing the exchange of sodium with calcium or with various singly-charged-cation neurotransmitters: substances that transmit impulses from one nerve cell to another and from a nerve cell to a muscle cell. Lithium has been shown to interact with the sodium-calcium exchange system in the plasma membranes of nerve and muscle cells and with the systems that take up neurotransmitters in the brain. An inherited defect in such a transport system in the brain might contribute to the abnormal state of brain and mind called mania.

I should emphasize that any discussion of the relation between the genetic determination of lithium-sodium countertransport and that of mania, or of the mechanism of the therapeutic action of lithium in mania, is highly speculative given the present state of ignorance. It is not known whether mania is the manifestation of an abnormal function of a small group of cells in the brain or whether it is a generalized disorder. It is also not known whether lithium improves mania by changing the function of a small group of brain cells or of many or all of them. The therapeutic action of lithium does not seem to be related to the variations among individuals in the maximum rate of lithium-sodium countertransport in red blood cells; it is equally effective in patients with an active countertransport system and in patients lacking a countertransport system.

Given the physicochemical similarities between lithium ions and the physiologically important, naturally occurring singly charged cations sodium and potassium (and to a lesser extent the doubly charged cations magnesium and calcium), it is not surprising that lithium at sufficiently high concentrations interacts with virtually all components of a living organism: proteins (including enzymes), nucleic acids and the charged groups of atoms on lipids and carbohydrates. It is also not surprising that the list of reported biological effects of lithium is quite long.

Even if attention is limited to the effects of concentrations of lithium in the therapeutic range, there are many promising avenues for the exploration of the mechanism of the ion's action in mania. Among the more attractive ones for me are (1) its effects on the synthesis of prostaglandins, (2) its effects on the production of white blood cells, (3) its effects on the transport of neurotransmitters and the sensitivity of receptors to them, (4) its inhibition of the choline transport system during the maturation of human red cells, (5) its action of in-

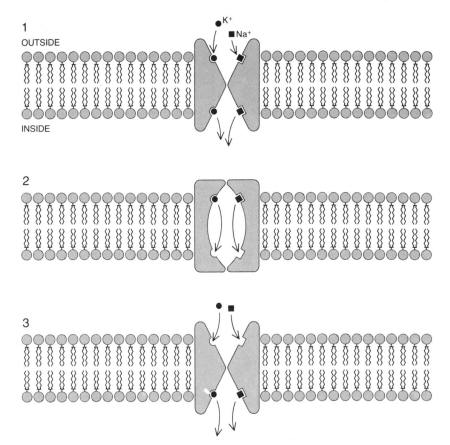

COTRANSPORT SYSTEM moves two kinds of ions simultaneously into or out of a membrane. In this hypothetical two-state, four-site model sodium and potassium ions move downhill into a red blood cell. With the membrane-penetrating protein open to the outer and inner surfaces of the membrane (*1*) a sodium ion and a potassium ion move from the blood plasma to binding sites near the outside opening of the protein and two similar ions move from interior binding sites to the inside of the cell. The configuration changes to a closed state (*2*) and the two ions from the blood plasma move to the inward-facing binding sites. They move into the cell (*3*) as the cycle starts again. In contrast to the models depicted for the sodium-potassium pump and for sodium-sodium exchange, sites can be empty in the closed state of this system.

creasing the concentration of choline in red cells, as recently reported by Ehrlich and Diamond, and (6) its inhibition of the activation of adenylate cyclase (the enzyme that catalyzes the action of intracellular "messenger" adenosine monophosphate, or AMP) by catecholamines, thyroid-stimulating hormone and antidiuretic hormone.

In view of the many sites of action of lithium, it is not surprising that it has many toxic effects. The safe range of therapeutic doses is fairly narrow. Among the more serious complications of treatment with lithium is that it can cause the kidneys to fail to respond to the antidiuretic hormone, giving rise to the disorder diabetes insipidus.

What does this incomplete chapter on lithium, membranes and mania portend for the future of medicine? It is another example of the increasing precision with which the physicochemical di-

mensions of disease can be determined. Only 30 years have passed since the first example of that kind of definition, the description of sickle-cell anemia as a molecular disease by Linus Pauling and Harvey A. Itano of the California Institute of Technology. Since then workers in many laboratories have done experiments that have identified not only the precise substitutions of amino acids taking place in sickle-cell anemia and other abnormal forms of hemoglobin but also the chemical identity of the genes directing the synthesis of this essential protein.

Similar perceptions are now emerging in the understanding of the molecular events in the development of such complex disorders as atherosclerosis and diabetes. For example, Joseph L. Goldstein and Michael S. Brown of the University of Texas Health Science Center at Dallas have identified the receptors responsible for the uptake by human fibroblast cells of low-density lipoproteins incorporating cholesterol and have shown that there is an inherited defect of these receptors in certain patients with a high level of cholesterol in their blood. Such patients are predisposed to develop atherosclerosis.

Similarly, it is now realized that diabetes mellitus (the common disorder characterized by increased concentrations of sugar in the urine and the blood plasma) has not one but many causes that may involve defects in the synthesis and processing of insulin or in its binding by receptors on cells and its subsequent action on the target cells. The defects may be inherited or may be due to an attack on the system by viruses.

All these advances suggest that the present classifying of disease into categories such as hypertension, diabetes, cancer and mania is crude. Classifications of this broad kind are in fact being supplanted by much more precise identifications of the molecular defects that underlie those general manifestations of physiological disorder. It is also being recognized in increasing detail how such molecular defects result from the interaction of genes and specific conditions associated with the environment. Rational and effective strategies for the prevention and control of complex diseases will be built on the basis of this kind of understanding.

To me the special appeal of the story of lithium, membranes and mania is its stark contrasts. It is somehow surprising and fascinating that a simple salt, an ion, an extract of rock, is able to alter such an ephemeral and subtle property of mind as mood. People are more accustomed to the idea that states of feeling are affected by relatively complex organic substances such as opium, marijuana, cocaine and alcohol. The physicochemical simplicity of lithium arouses the hope that it will provide a light to clarify the neuronal basis of mood.

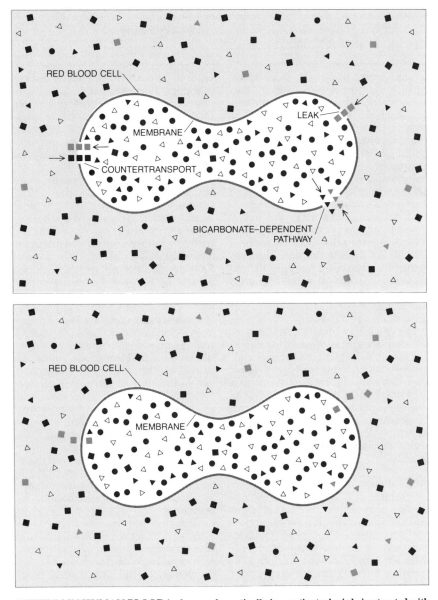

LITHIUM IN HUMAN BLOOD is shown schematically in a patient who is being treated with a lithium salt. The ions that figure in the process are lithium (*colored rectangles*), sodium (*black rectangles*), potassium (*black circles*), lithium carbonate (*colored triangles*), bicarbonate (*black triangles*) and chloride (*open triangles*). The chemical symbols for these cations and anions are respectively Li^+, Na^+ and K^+, $LiCO_3^-$, HCO_3^- and Cl^-. In a patient under lithium therapy the concentration of Li^+ in the blood plasma (*light color*) is held at about one millimolar. The concentration inside a cell is kept at a lower value (about .3 millimolar) by the operation of a lithium-sodium countertransport system. When a steady state is reached, the outward movement of lithium through the countertransport system is equaled by inward movement of the ions through two pathways: a leak and an anion-exchange pathway that exchanges $LiCO_3^-$ for HCO_3^- or Cl^-. Several movements through the membrane are shown at the top; the steady-state situation at any given moment is shown at the bottom. In each representation the relative proportions of the various cations and anions in the cell and in the plasma are indicated.

12

The Mind-Body Problem

by Jerry A. Fodor
January 1981

*Could calculating machines have pains, Martians have
expectations and disembodied spirits have thoughts?
The modern functionalist approach to psychology raises
the logical possibility that they could*

Modern philosophy of science has been devoted largely to the formal and systematic description of the successful practices of working scientists. The philosopher does not try to dictate how scientific inquiry and argument ought to be conducted. Instead he tries to enumerate the principles and practices that have contributed to good science. The philosopher has devoted the most attention to analyzing the methodological peculiarities of the physical sciences. The analysis has helped to clarify the nature of confirmation, the logical structure of scientific theories, the formal properties of statements that express laws and the question of whether theoretical entities actually exist.

It is only rather recently that philosophers have become seriously interested in the methodological tenets of psychology. Psychological explanations of behavior refer liberally to the mind and to states, operations and processes of the mind. The philosophical difficulty comes in stating in unambiguous language what such references imply.

Traditional philosophies of mind can be divided into two broad categories: dualist theories and materialist theories. In the dualist approach the mind is a nonphysical substance. In materialist theories the mental is not distinct from the physical; indeed, all mental states, properties, processes and operations are in principle identical with physical states, properties, processes and operations. Some materialists, known as behaviorists, maintain that all talk of mental causes can be eliminated from the language of psychology in favor of talk of environmental stimuli and behavioral responses. Other materialists, the identity theorists, contend that there are mental causes and that they are identical with neurophysiological events in the brain.

In the past 15 years a philosophy of mind called functionalism that is neither dualist nor materialist has emerged from philosophical reflection on developments in artificial intelligence, computational theory, linguistics, cybernetics and psychology. All these fields, which are collectively known as the cognitive sciences, have in common a certain level of abstraction and a concern with systems that process information. Functionalism, which seeks to provide a philosophical account of this level of abstraction, recognizes the possibility that systems as diverse as human beings, calculating machines and disembodied spirits could all have mental states. In the functionalist view the psychology of a system depends not on the stuff it is made of (living cells, metal or spiritual energy) but on how the stuff is put together. Functionalism is a difficult concept, and one way of coming to grips with it is to review the deficiencies of the dualist and materialist philosophies of mind it aims to displace.

The chief drawback of dualism is its failure to account adequately for mental causation. If the mind is nonphysical, it has no position in physical space. How, then, can a mental cause give rise to a behavioral effect that has a position in space? To put it another way, how can the nonphysical give rise to the physical without violating the laws of the conservation of mass, of energy and of momentum?

The dualist might respond that the problem of how an immaterial substance can cause physical events is not much obscurer than the problem of how one physical event can cause another. Yet there is an important difference: there are many clear cases of physical causation but not one clear case of nonphysical causation. Physical interaction is something philosophers, like all other people, have to live with. Nonphysical interaction, however, may be no more than an artifact of the immaterialist construal of the mental. Most philosophers now agree that no argument has successfully demonstrated why mind-body causation should not be regarded as a species of physical causation.

Dualism is also incompatible with the practices of working psychologists. The psychologist frequently applies the experimental methods of the physical sciences to the study of the mind. If mental processes were different in kind from physical processes, there would be no reason to expect these methods to work in the realm of the mental. In order to justify their experimental methods many psychologists urgently sought an alternative to dualism.

In the 1920's John B. Watson of Johns Hopkins University made the radical suggestion that behavior does not have mental causes. He regarded the behavior of an organism as its observable responses to stimuli, which he took to be the causes of its behavior. Over the next 30 years psychologists such as B. F. Skinner of Harvard University developed Watson's ideas into an elaborate world view in which the role of psychology was to catalogue the laws that determine causal relations between stimuli and responses. In this "radical behaviorist" view the problem of explaining the nature of the mind-body interaction vanishes; there is no such interaction.

Radical behaviorism has always worn an air of paradox. For better or worse, the idea of mental causation is deeply ingrained in our everyday language and in our ways of understanding our fellow men and ourselves. For example, people commonly attribute behavior to beliefs, to knowledge and to expectations. Brown puts gas in his tank because he believes the car will not run without it. Jones writes not "acheive" but "achieve" because he knows the rule about putting *i* before *e*. Even when a behavioral response is closely tied to an environmental stimulus, mental processes often intervene. Smith carries an umbrella because the sky is cloudy, but the weather is only part of the story. There are apparently also mental links in the causal chain: observation and expectation. The clouds affect Smith's behavior only because he observes them and because they induce in him an expectation of rain.

The radical behaviorist is unmoved by appeals to such cases. He is prepared to dismiss references to mental causes, however plausible they may seem, as the residue of outworn creeds. The radical behaviorist predicts that as psychologists come to understand more about the relations between stimuli and responses they will find it increasingly possible to explain behavior without postulating mental causes.

The strongest argument against behaviorism is that psychology has not turned out this way; the opposite has happened. As psychology has matured, the framework of mental states and processes that is apparently needed to account for experimental observations has grown all the more elaborate. Particularly in the case of human behavior psychological theories satisfying the methodological tenets of radical behaviorism have proved largely sterile, as would be expected if the postulated mental processes are real and causally effective.

Nevertheless, many philosophers were initially drawn to radical behaviorism because, paradoxes and all, it seemed better than dualism. Since a psychology committed to immaterial substances was unacceptable, philosophers turned to radical behaviorism because it seemed to be the only alternative materialist philosophy of mind. The choice, as they saw it, was between radical behaviorism and ghosts.

By the early 1960's philosophers began to have doubts that dualism and radical behaviorism exhausted the possible approaches to the philosophy of mind. Since the two theories seemed unattractive, the right strategy might be to develop a materialist philosophy of mind that nonetheless allowed for mental causes. Two such philosophies emerged, one called logical behaviorism and the other called the central-state identity theory.

Logical behaviorism is a semantic theory about what mental terms mean. The basic idea is that attributing a mental state (say thirst) to an organism is the same as saying that the organism is disposed to behave in a particular way (for example to drink if there is water available). On this view every mental ascription is equivalent in meaning to an if-then statement (called a behavioral hypothetical) that expresses a behavioral disposition. For example, "Smith is thirsty" might be taken to be equivalent to the dispositional statement "If there were water available, then Smith would drink some." By definition a behavioral hypothetical includes no mental terms. The if-clause of the hypothetical speaks only of stimuli and the then-clause speaks only of behavioral responses. Since stimuli and responses are physical events, logical behaviorism is a species of materialism.

The strength of logical behaviorism is that by translating mental language into the language of stimuli and responses it provides an interpretation of psychological explanations in which behavioral

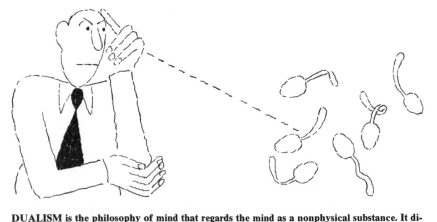

DUALISM is the philosophy of mind that regards the mind as a nonphysical substance. It divides everything there is in the world into two distinct categories: the mental and the physical. The chief difficulty with dualism is its failure to account adequately for the causal interaction of the mental and the physical. It is not evident how a nonphysical mind could give rise to any physical effects without violating the laws of conservation of mass, energy and momentum.

effects are attributed to mental causes. Mental causation is simply the manifestation of a behavioral disposition. More precisely, mental causation is what happens when an organism has a behavioral disposition and the if-clause of the behavioral hypothetical expressing the disposition happens to be true. For example, the causal statement "Smith drank some water because he was thirsty" might be taken to mean "If there were

water available, then Smith would drink some, and there was water available."

I have somewhat oversimplified logical behaviorism by assuming that each mental ascription can be translated by a unique behavioral hypothetical. Actually the logical behaviorist often maintains that it takes an open-ended set (perhaps an infinite set) of behavioral hypotheticals to spell out the behavioral disposition expressed by a mental term.

RADICAL BEHAVIORISM is the philosophy of mind that denies the existence of the mind and mental states, properties, processes and operations. The radical behaviorist believes behavior does not have mental causes. He considers the behavior of an organism to be its responses to stimuli. The role of psychology is to catalogue the relations between stimuli and responses.

The mental ascription "Smith is thirsty" might also be satisfied by the hypothetical "If there were orange juice available, then Smith would drink some" and by a host of other hypotheticals. In any event the logical behaviorist does not usually maintain he can actually enumerate all the hypotheticals that correspond to a behavioral disposition expressing a given mental term. He only insists that in principle the meaning of any mental term can be conveyed by behavioral hypotheticals.

The way the logical behaviorist has interpreted a mental term such as thirsty is modeled after the way many philosophers have interpreted a physical disposition such as fragility. The physical disposition "The glass is fragile" is often taken to mean something like "If the glass were struck, then it would break." By the same token the logical behaviorist's analysis of mental causation is similar to the received analysis of one kind of physical causation. The causal statement "The glass broke because it was fragile" is taken to mean something like "If the glass were struck, then it would break, and the glass was struck."

By equating mental terms with behavioral dispositions the logical behaviorist has put mental terms on a par with the nonbehavioral dispositions of the physical sciences. That is a promising move, because the analysis of nonbehavioral dispositions is on relatively solid philosophical ground. An explanation attributing the breaking of a glass to its fragility is surely something even the staunchest materialist can accept. By arguing that mental terms are synonymous with dispositional terms, the logical behaviorist has provided something the radical behaviorist could not: a materialist account of mental causation.

Nevertheless, the analogy between mental causation as construed by the logical behaviorist and physical causation goes only so far. The logical behaviorist treats the manifestation of a disposition as the sole form of mental causation, whereas the physical sciences recognize additional kinds of causation. There is the kind of causation where one physical event causes another, as when the breaking of a glass is attributed to its having been struck. In fact, explanations that involve event-event causation are presumably more basic than dispositional explanations, because the manifestation of a disposition (the breaking of a fragile glass) always involves event-event causation and not vice versa. In the realm of the mental many examples of event-event causation involve one mental state's causing another, and for this kind of causation logical behaviorism provides no analysis. As a result the logical behaviorist is committed to the tacit and implausible assumption that psychology requires a less robust notion of causation than the physical sciences require.

Event-event causation actually seems to be quite common in the realm of the mental. Mental causes typically give rise to behavioral effects by virtue of their interaction with other mental causes. For example, having a headache causes a disposition to take aspirin only if one also has the desire to get rid of the headache, the belief that aspirin exists, the belief that taking aspirin reduces headaches and so on. Since mental states interact in generating behavior, it will be necessary to find a construal of psychological explanations that posits mental processes: causal sequences of mental events. It is this construal that logical behaviorism fails to provide.

Such considerations bring out a fundamental way in which logical behaviorism is quite similar to radical behaviorism. It is true that the logical behaviorist, unlike the radical behaviorist, acknowledges the existence of mental states. Yet since the underlying tenet of logical behaviorism is that references to mental states can be translated out of psychological explanations by employing behavioral hypotheticals, all talk of mental states and processes is in a sense heuristic. The only facts to which the behaviorist is actually committed are facts about relations between stimuli and responses. In this respect logical behaviorism is just radical behaviorism in a semantic form. Although the former theory offers a construal of mental causation, the construal is Pickwickian. What does not really exist cannot cause anything, and the logical behaviorist, like the radical behaviorist, believes deep down that mental causes do not exist.

An alternative materialist theory of the mind to logical behaviorism is the central-state identity theory. According to this theory, mental events, states and processes are identical with neurophysiological events in the brain, and the property of being in a certain mental state (such as having a headache or believing it will rain) is identical with the property of being in a certain neurophysiological state. On this basis it is easy to make sense of the idea that a behavioral effect might sometimes have a chain of mental causes; that will be the case whenever a behavioral effect is contingent on the appropriate sequence of neurophysiological events.

The central-state identity theory acknowledges that it is possible for mental

LOGICAL BEHAVIORISM is a semantic thesis about what mental terms mean. The logical behaviorist maintains that mental terms express behavioral dispositions. Consider the mental state of being thirsty. The logical behaviorist maintains that the sentence "Smith is thirsty" might be taken as equivalent in meaning to the dispositional statement "If there were water available, then Smith would drink some." The strength of logical behaviorism is that it provides an account of mental causation: the realization of a behavioral disposition. For example, the causal statement "Smith drank some water because he was thirsty" might be taken to mean "If there were water available, then Smith would drink some, and there was water available."

CENTRAL-STATE IDENTITY THEORY is the philosophy of mind that equates mental events, states and processes with neuro- **physiological events. Property of being in a given mental state is identical with the property of being in a given neurophysiological state.**

causes to interact causally without ever giving rise to any behavioral effect, as when a person thinks for a while about what he ought to do and then decides to do nothing. If mental processes are neurophysiological, they must have the causal properties of neurophysiological processes. Since neurophysiological processes are presumably physical processes, the central-state identity theory ensures that the concept of mental causation is as rich as the concept of physical causation.

The central-state identity theory provides a satisfactory account of what the mental terms in psychological explanations refer to, and so it is favored by psychologists who are dissatisfied with behaviorism. The behaviorist maintains that mental terms refer to nothing or that they refer to the parameters of stimulus-response relations. Either way the existence of mental entities is only illusory. The identity theorist, on the other hand, argues that mental terms refer to neurophysiological states. Thus he can take seriously the project of explaining behavior by appealing to its mental causes.

The chief advantage of the identity theory is that it takes the explanatory constructs of psychology at face value, which is surely something a philosophy of mind ought to do if it can. The identi-

ty theory shows how the mentalistic explanations of psychology could be not mere heuristics but literal accounts of the causal history of behavior. Moreover, since the identity theory is not a semantic thesis, it is immune to many arguments that cast in doubt logical behaviorism. A drawback of logical behaviorism is that the observation "John has a headache" does not seem to mean the same thing as a statement of the form "John is disposed to behave in such and such a way." The identity theorist, however, can live with the fact that "John has a headache" and "John is in such and such a brain state" are not synonymous. The assertion of the identity theorist is not that these sentences mean the same thing but only that they are rendered true (or false) by the same neurophysiological phenomena.

The identity theory can be held either as a doctrine about mental particulars (John's current pain or Bill's fear of animals) or as a doctrine about mental universals, or properties (having a pain or being afraid of animals). The two doctrines, called respectively token physicalism and type physicalism, differ in strength and plausibility. Token physicalism maintains only that all the mental particulars that happen to exist are neurophysiological, whereas type physicalism makes the more sweeping asser-

tion that all the mental particulars there could possibly be are neurophysiological. Token physicalism does not rule out the logical possibility of machines and disembodied spirits having mental properties. Type physicalism dismisses this possibility because neither machines nor disembodied spirits have neurons.

Type physicalism is not a plausible doctrine about mental properties even if token physicalism is right about mental particulars. The problem with type physicalism is that the psychological constitution of a system seems to depend not on its hardware, or physical composition, but on its software, or program. Why should the philosopher dismiss the possibility that silicon-based Martians have pains, assuming that the silicon is properly organized? And why should the philosopher rule out the possibility of machines having beliefs, assuming that the machines are correctly programmed? If it is logically possible that Martians and machines could have mental properties, then mental properties and neurophysiological processes cannot be identical, however much they may prove to be coextensive.

What it all comes down to is that there seems to be a level of abstraction at which the generalizations of psychology are most naturally pitched. This level of abstraction cuts across differences in the physical composition of the systems to which psychological generalizations apply. In the cognitive sciences, at least, the natural domain for psychological theorizing seems to be all systems that process information. The problem with type physicalism is that there are possible information-processing systems with the same psychological constitution as human beings but not the same physical organization. In principle all kinds of physically different things could have human software.

This situation calls for a relational account of mental properties that abstracts them from the physical structure of their bearers. In spite of the objections to logical behaviorism that I presented above, logical behaviorism was at least on the right track in offering a relational interpretation of mental properties: to have a headache is to be disposed to exhibit a certain pattern of relations between the stimuli one encounters and the responses one exhibits. If that is what having a headache is, how-

ever, there is no reason in principle why only heads that are physically similar to ours can ache. Indeed, according to logical behaviorism, it is a necessary truth that any system that has our stimulus-response contingencies also has our headaches.

All of this emerged 10 or 15 years ago as a nasty dilemma for the materialist program in the philosophy of mind. On the one hand the identity theorist (and not the logical behaviorist) had got right the causal character of the interactions of mind and body. On the other the logical behaviorist (and not the identity theorist) had got right the relational character of mental properties. Functionalism has apparently been able to resolve the dilemma. By stressing the distinction computer science draws between hardware and software the functionalist can make sense of both the causal and the relational character of the mental.

The intuition underlying functionalism is that what determines the psychological type to which a mental particular belongs is the causal role of the particu-

lar in the mental life of the organism. Functional individuation is differentiation with respect to causal role. A headache, for example, is identified with the type of mental state that among other things causes a disposition for taking aspirin in people who believe aspirin relieves a headache, causes a desire to rid oneself of the pain one is feeling, often causes someone who speaks English to say such things as "I have a headache" and is brought on by overwork, eyestrain and tension. This list is presumably not complete. More will be known about the nature of a headache as psychological and physiological research discovers more about its causal role.

Functionalism construes the concept of causal role in such a way that a mental state can be defined by its causal relations to other mental states. In this respect functionalism is completely different from logical behaviorism. Another major difference is that functionalism is not a reductionist thesis. It does not foresee, even in principle,

FUNCTIONALISM is the philosophy of mind based on the distinction that computer science draws between a system's hardware, or physical composition, and its software, or program. The psychology of a system such as a human being, a machine or a disembodied spirit does not depend on the stuff the system is made of (neurons, diodes or spiritual energy) but on how that stuff is organized. Functionalism does not rule out the possibility, however remote it may be, of mechanical and ethereal systems having mental states and processes.

the elimination of mentalistic concepts from the explanatory apparatus of psychological theories.

The difference between functionalism and logical behaviorism is brought out by the fact that functionalism is fully compatible with token physicalism. The functionalist would not be disturbed if brain events turn out to be the only things with the functional properties that define mental states. Indeed, most functionalists fully expect it will turn out that way.

Since functionalism recognizes that mental particulars may be physical, it is compatible with the idea that mental causation is a species of physical causation. In other words, functionalism tolerates the materialist solution to the mind-body problem provided by the central-state identity theory. It is possible for the functionalist to assert both that mental properties are typically defined in terms of their relations and that interactions of mind and body are typically causal in however robust a notion of causality is required by psychological explanations. The logical behaviorist can endorse only the first assertion and the type physicalist only the second. As a result functionalism seems to capture the best features of the materialist alternatives to dualism. It is no wonder that functionalism has become increasingly popular.

Machines provide good examples of two concepts that are central to functionalism: the concept that mental states are interdefined and the concept that they can be realized by many systems. The illustration on the next page contrasts a behavioristic Coke machine with a mentalistic one. Both machines dispense a Coke for 10 cents. (The price has not been affected by inflation.) The states of the machines are defined by reference to their causal roles, but only the machine on the left would satisfy the behaviorist. Its single state ($S0$) is completely specified in terms of stimuli and responses. $S0$ is the state a machine is in if, and only if, given a dime as the input, it dispenses a Coke as the output.

The machine on the right in the illustration has interdefined states ($S1$ and $S2$), which are characteristic of functionalism. $S1$ is the state a machine is in if, and only if, (1) given a nickel, it dispenses nothing and proceeds to $S2$, and (2) given a dime, it dispenses a Coke and stays in $S1$. $S2$ is the state a machine is in if, and only if, (1) given a nickel, it dispenses a Coke and proceeds to $S1$, and (2) given a dime, it dispenses a Coke and a nickel and proceeds to $S1$. What $S1$ and $S2$ jointly amount to is the machine's dispensing a Coke if it is given a dime, dispensing a Coke and a nickel if it is given a dime and a nickel and waiting to be given a second nickel if it has been given a first one.

Since $S1$ and $S2$ are each defined by hypothetical statements, they can be viewed as dispositions. Nevertheless, they are not behavioral dispositions because the consequences an input has for a machine in $S1$ or $S2$ are not specified solely in terms of the output of the machine. Rather, the consequences also involve the machine's internal states.

Nothing about the way I have described the behavioristic and mentalistic Coke machines puts constraints on what they could be made of. Any system whose states bore the proper relations to inputs, outputs and other states could be one of these machines. No doubt it is reasonable to expect such a system to be constructed out of such things as wheels, levers and diodes (token physicalism for Coke machines). Similarly, it is reasonable to expect that our minds may prove to be neurophysiological (token physicalism for human beings).

Nevertheless, the software description of a Coke machine does not logically require wheels, levers and diodes for its concrete realization. By the same token, the software description of the mind does not logically require neurons. As far as functionalism is concerned a Coke machine with states $S1$ and $S2$ could be made of ectoplasm, if there is such stuff and if its states have the right causal properties. Functionalism allows for the possibility of disembodied Coke machines in exactly the same way and to the same extent that it allows for the possibility of disembodied minds.

To say that $S1$ and $S2$ are interdefined and realizable by different kinds of hardware is not, of course, to say that a Coke machine has a mind. Although interdefinition and functional specification are typical features of mental states, they are clearly not sufficient for mentality. What more is required is a question to which I shall return below.

Some philosophers are suspicious of functionalism because it seems too easy. Since functionalism licenses the individuation of states by reference to their causal role, it appears to allow a trivial explanation of any observed event E, that is, it appears to postulate an E-causer. For example, what makes the valves in a machine open? Why, the operation of a valve opener. And what is a valve opener? Why, anything that has the functionally defined property of causing valves to open.

In psychology this kind of question-begging often takes the form of theories that in effect postulate homunculi with the selfsame intellectual capacities the theorist set out to explain. Such is the case when visual perception is explained by simply postulating psychological mechanisms that process visual information. The behaviorist has often charged the mentalist, sometimes justifiably, of mongering this kind of question-begging pseudo explanation. The

charge will have to be met if functionally defined mental states are to have a serious role in psychological theories.

The burden of the accusation is not untruth but triviality. There can be no doubt that it is a valve opener that opens valves, and it is likely that visual perception is mediated by the processing of visual information. The charge is that such putative functional explanations are mere platitudes. The functionalist can meet this objection by allowing functionally defined theoretical constructs only where mechanisms exist that can carry out the function and only where he has some notion of what such mechanisms might be like. One way of imposing this requirement is to identify the mental processes that psychology postulates with the operations of the restricted class of possible computers called Turing machines.

A Turing machine can be informally characterized as a mechanism with a finite number of program states. The inputs and outputs of the machine are written on a tape that is divided into squares each of which includes a symbol from a finite alphabet. The machine scans the tape one square at a time. It can erase the symbol on a scanned square and print a new one in its place. The machine can execute only the elementary mechanical operations of scanning, erasing, printing, moving the tape and changing state.

The program states of the Turing machine are defined solely in terms of the input symbols on the tape, the output symbols on the tape, the elementary operations and the other states of the program. Each program state is therefore functionally defined by the part it plays in the overall operation of the machine. Since the functional role of a state depends on the relation of the state to other states as well as to inputs and outputs, the relational character of the mental is captured by the Turing-machine version of functionalism. Since the definition of a program state never refers to the physical structure of the system running the program, the Turing-machine version of functionalism also captures the idea that the character of a mental state is independent of its physical realization. A human being, a roomful of people, a computer and a disembodied spirit would all be a Turing machine if they operated according to a Turing-machine program.

The proposal is to restrict the functional definition of psychological states to those that can be expressed in terms of the program states of Turing machines. If this restriction can be enforced, it provides a guarantee that psychological theories will be compatible with the demands of mechanisms. Since Turing machines are very simple devices, they are in principle quite easy to build. Consequently by formulating a psychological explanation as a Turing-machine program the psychologist ensures that the explanation is mechanistic, even though the hardware realizing the mechanism is left open.

There are many kinds of computational mechanisms other than Turing machines, and so the formulation of a functionalist psychological theory in Turing-machine notation provides only a sufficient condition for the theory's being mechanically realizable. What makes the condition interesting, however, is that the simple Turing machine can perform many complex tasks. Although the elementary operations of the Turing machine are restricted, iterations of the operations enable the machine to carry out any well-defined computation on discrete symbols.

An important tendency in the cognitive sciences is to treat the mind chiefly as a device that manipulates symbols. If

a mental process can be functionally defined as an operation on symbols, there is a Turing machine capable of carrying out the computation and a variety of mechanisms for realizing the Turing machine. Where the manipulation of symbols is important the Turing machine provides a connection between functional explanation and mechanistic explanation.

The reduction of a psychological theory to a program for a Turing machine is a way of exorcising the homunculi. The reduction ensures that no operations have been postulated except those that could be performed by a familiar mechanism. Of course, the working psychologist usually cannot specify the reduction for each functionally individuated process in every theory he is prepared to take seriously. In practice the argument usually goes in the opposite direction; if the postulation of a mental operation is essential to some cherished

psychological explanation, the theorist tends to assume that there must be a program for a Turing machine that will carry out that operation.

The "black boxes" that are common in flow charts drawn by psychologists often serve to indicate postulated mental processes for which Turing reductions are wanting. Even so, the possibility in principle of such reductions serves as a methodological constraint on psychological theorizing by determining what functional definitions are to be allowed and what it would be like to know that everything has been explained that could possibly need explanation.

Such is the origin, the provenance and the promise of contemporary functionalism. How much has it actually paid off? This question is not easy to answer because much of what is now happening in the philosophy of mind and the cognitive sciences is directed at exploring the

	STATE SO
DIME INPUT	DISPENSES A COKE

	STATE S1	STATE S2
NICKEL INPUT	GIVES NO OUTPUT AND GOES TO S2	DISPENSES A COKE AND GOES TO S1
DIME INPUT	DISPENSES A COKE AND STAYS IN S1	DISPENSES A COKE AND A NICKEL AND GOES TO S1

TWO COKE MACHINES bring out the difference between behaviorism (the doctrine that there are no mental causes) and mentalism (the doctrine that there are mental causes). Both machines dispense a Coke for 10 cents and have states that are defined by reference to their causal role. The machine at the left is a behavioristic one: its single state ($S0$) is defined solely in terms of the input and the output. The machine at the right is a mentalistic one: its two states ($S1$, $S2$) must be defined not only in terms of the input and the output but also in terms of each other. To put it another way, the output of the Coke machine depends on the state the machine is in as well as on the input. The functionalist philosopher maintains that mental states are interdefined, like the internal states of the mentalistic Coke machine.

scope and limits of the functionalist explanations of behavior. I shall, however, give a brief overview.

An obvious objection to functionalism as a theory of the mind is that the functionalist definition is not limited to mental states and processes. Catalysts, Coke machines, valve openers, pencil sharpeners, mousetraps and ministers of finance are all in one way or another concepts that are functionally defined, but none is a mental concept such as pain, belief and desire. What, then, characterizes the mental? And can it be captured in a functionalist framework?

The traditional view in the philosophy of mind has it that mental states are distinguished by their having what are called either qualitative content or intentional content. I shall discuss qualitative content first.

It is not easy to say what qualitative content is; indeed, according to some theories, it is not even possible to say what it is because it can be known not by description but only by direct experience. I shall nonetheless attempt to describe it. Try to imagine looking at a blank wall through a red filter. Now change the filter to a green one and leave everything else exactly the way it was. Something about the character of your experience changes when the filter does, and it is this kind of thing that philosophers call qualitative content. I am not entirely comfortable about introducing qualitative content in this way, but it is a subject with which many philosophers are not comfortable.

The reason qualitative content is a problem for functionalism is straightforward. Functionalism is committed to defining mental states in terms of their causes and effects. It seems, however, as if two mental states could have all the same causal relations and yet could differ in their qualitative content. Let me illustrate this with the classic puzzle of the inverted spectrum.

It seems possible to imagine two observers who are alike in all relevant psychological respects except that experiences having the qualitative content of red for one observer would have the qualitative content of green for the other. Nothing about their behavior need reveal the difference because both of them see ripe tomatoes and flaming sunsets as being similar in color and both of them call that color "red." Moreover, the causal connection between their (qualitatively distinct) experiences and their other mental states could also be identical. Perhaps they both think of Little Red Riding Hood when they see ripe tomatoes, feel depressed when they see the color green and so on. It seems as if anything that could be packed into the notion of the causal role of their experiences could be shared by them, and yet the qualitative content of the experiences could be as different as you like. If this is possible, then the functionalist account does not work for mental states that have qualitative content. If one person is having a green experience while another person is having a red one, then surely they must be in different mental states.

The example of the inverted spectrum is more than a verbal puzzle. Having qualitative content is supposed to be a chief factor in what makes a mental state conscious. Many psychologists who are inclined to accept the functionalist framework are nonetheless worried about the failure of functionalism to reveal much about the nature of consciousness. Functionalists have made a few ingenious attempts to talk themselves and their colleagues out of this worry, but they have not, in my view, done so with much success. (For example, perhaps one is wrong in thinking one can imagine what an inverted spectrum would be like.) As matters stand, the problem of qualitative content poses a serious threat to the assertion that functionalism can provide a general theory of the mental.

Functionalism has fared much better with the intentional content of mental states. Indeed, it is here that the major achievements of recent cognitive science are found. To say that a mental state has intentional content is to say that it has certain semantic properties. For example, for Enrico to believe Galileo was Italian apparently involves a three-way relation between Enrico, a belief and a proposition that is the content of the belief (namely the proposition that Galileo was Italian). In particular it is an essential property of Enrico's belief that it is about Galileo (and not about, say, Newton) and that it is true if, and only if, Galileo was indeed Italian. Philosophers are divided on how these considerations fit together, but it is widely agreed that beliefs involve semantic properties such as expressing a proposition, being true or false and being about one thing rather than another.

It is important to understand the semantic properties of beliefs because theories in the cognitive sciences are largely about the beliefs organisms have. Theories of learning and perception, for example, are chiefly accounts of how the host of beliefs an organism has are determined by the character of its experiences and its genetic endowment. The functionalist account of mental states does not by itself provide the required insights. Mousetraps are functionally defined, yet mousetraps do not express propositions and they are not true or false.

There is at least one kind of thing other than a mental state that has intentional content: a symbol. Like thoughts, symbols seem to be about things. If someone says "Galileo was Italian," his utterance, like Enrico's belief, expresses a proposition about Galileo that is true or false depending on Galileo's homeland. This parallel between the symbolic and the mental underlies the traditional quest for a unified treatment of language and mind. Cognitive science is now trying to provide such a treatment.

The basic concept is simple but striking. Assume that there are such things as mental symbols (mental representations) and that mental symbols have semantic properties. On this view having a belief involves being related to a mental symbol, and the belief inherits its semantic properties from the mental symbol that figures in the relation. Mental processes (thinking, perceiving, learning and so on) involve causal interactions among relational states such as having a belief. The semantic properties of the words and sentences we utter are in turn inherited from the semantic properties of the mental states that language expresses.

Associating the semantic properties of mental states with those of mental symbols is fully compatible with the computer metaphor, because it is natural to think of the computer as a mechanism that manipulates symbols. A computation is a causal chain of computer states and the links in the chain are operations on semantically interpreted formulas in a machine code. To think of a system (such as the nervous system) as a computer is to raise questions about the nature of the code in which it computes and the semantic properties of the symbols in the code. In fact, the analogy between minds and computers actually implies the postulation of mental symbols. There is no computation without representation.

The representational account of the mind, however, predates considerably the invention of the computing machine. It is a throwback to classical epistemology, which is a tradition that includes philosophers as diverse as John Locke, David Hume, George Berkeley, René Descartes, Immanuel Kant, John Stuart Mill and William James.

Hume, for one, developed a representational theory of the mind that included five points. First, there exist "Ideas," which are a species of mental symbol. Second, having a belief involves entertaining an Idea. Third, mental processes are causal associations of Ideas. Fourth, Ideas are like pictures. And fifth, Ideas have their semantic properties by virtue of what they resemble: the Idea of John is about John because it looks like him.

Contemporary cognitive psychologists do not accept the details of Hume's theory, although they endorse much of its spirit. Theories of computation provide a far richer account of mental processes than the mere association of Ideas. And only a few psychologists still think that imagery is the chief vehicle of mental representation. Nevertheless, the most significant break with Hume's theory lies in the abandoning of resemblance as an explanation of the semantic properties of mental representations.

Many philosophers, starting with Berkeley, have argued that there is something seriously wrong with the suggestion that the semantic relation between a thought and what the thought is about could be one of resemblance. Consider the thought that John is tall. Clearly the thought is true only of the state of affairs consisting of John's being tall. A theory of the semantic properties of a thought should therefore explain how this particular thought is related to this particular state of affairs. According to the resemblance theory, entertaining the thought involves having a mental image that shows John to be tall. To put it another way, the relation between the thought that John is tall and his being tall is like the relation between a tall man and his portrait.

The difficulty with the resemblance theory is that any portrait showing John to be tall must also show him to be many other things: clothed or naked, lying, standing or sitting, having a head or not having one, and so on. A portrait of a tall man who is sitting down resembles a man's being seated as much as it resembles a man's being tall. On the resemblance theory it is not clear what distinguishes thoughts about John's height from thoughts about his posture.

The resemblance theory turns out to encounter paradoxes at every turn. The possibility of construing beliefs as involving relations to semantically interpreted mental representations clearly depends on having an acceptable account of where the semantic properties of the mental representations come from. If resemblance will not provide this account, what will?

The current idea is that the semantic properties of a mental representation are determined by aspects of its functional role. In other words, a sufficient condition for having semantic properties can be specified in causal terms. This is the connection between functionalism and the representational theory of the mind. Modern cognitive psychology rests largely on the hope that these two doctrines can be made to support each other.

No philosopher is now prepared to say exactly how the functional role of a mental representation determines its semantic properties. Nevertheless, the functionalist recognizes three types of causal relation among psychological states involving mental representations, and they might serve to fix the semantic properties of mental representations. The three types are causal relations among mental states and stimuli, mental states and responses and some mental states and other ones.

Consider the belief that John is tall. Presumably the following facts, which correspond respectively to the three types of causal relation, are relevant to determining the semantic properties of the mental representation involved in the belief. First, the belief is a normal effect of certain stimulations, such as seeing John in circumstances that reveal his height. Second, the belief is the normal cause of certain behavioral effects, such as uttering "John is tall." Third, the belief is a normal cause of certain other beliefs and a normal effect of certain other beliefs. For example, anyone who believes John is tall is very likely also to believe someone is tall. Having the first belief is normally causally sufficient for having the second belief. And anyone who believes everyone in the room is tall and also believes John is in the room will very likely believe John is tall. The third belief is a normal effect of the first two. In short, the functionalist maintains that the proposition expressed by a given mental representation depends on the causal properties of the mental states in which that mental representation figures.

The concept that the semantic properties of mental representations are determined by aspects of their functional role is at the center of current work in the cognitive sciences. Nevertheless, the concept may not be true. Many philosophers who are unsympathetic to the cognitive turn in modern psychology doubt its truth, and many psychologists would probably reject it in the bald and unelaborated way that I have sketched it. Yet even in its skeletal form, there is this much to be said in its favor: It legitimizes the notion of mental representation, which has become increasingly important to theorizing in every branch of the cognitive sciences. Recent advances in formulating and testing hypotheses about the character of mental representations in fields ranging from phonetics to computer vision suggest that the concept of mental representation is fundamental to empirical theories of the mind.

The behaviorist has rejected the appeal to mental representation because it runs counter to his view of the explanatory mechanisms that can figure in psychological theories. Nevertheless, the science of mental representation is now flourishing. The history of science reveals that when a successful theory comes into conflict with a methodological scruple, it is generally the scruple that gives way. Accordingly the functionalist has relaxed the behaviorist constraints on psychological explanations. There is probably no better way to decide what is methodologically permissible in science than by investigating what successful science requires.

THE AUTHORS

CHARLES F. STEVENS ("The Neuron") is professor of physiology and chairman of the section of molecular neurobiology at the Yale University School of Medicine. He did his undergraduate work in experimental psychology at Harvard University and obtained his M.D. at the Yale School of Medicine and his Ph.D. in biophysics from Rockefeller University in 1964. From 1963 through 1975 he was on the physiology and biophysics faculty of the University of Washington School of Medicine, taking a sabbatical leave during the 1969–70 academic year at the Lorentz Institute for Theoretical Physics at the University of Leiden.

RODOLFO R. LLINÁS is professor and chairman of the department of physiology and biophysics at the New York University Medical Center in New York. Born in Colombia, he received his M.D. from the Pontifical University of Javeriana in 1959. He then moved to Australia as a research scholar at the Australian National University, from which he got his Ph.D. in 1965. He came to the U.S. in the same year as associate professor at the University of Minnesota. From 1966 to 1970 he was on the staff of the Institute for Biomedical Research of the American Medical Association Education and Research Foundation. In 1970 he became professor of physiology and biophysics and head of the division of neurobiology at the University of Iowa; he left Iowa in 1976 to take up his present jobs at New York University. In addition to synaptic transmission Llinás has worked on the evolution of the central nervous system.

LESLIE L. IVERSEN ("The Chemistry of the Brain") is director of the Medical Research Council Pharmacology Unit at Cambridge in England. He did his undergraduate work at Trinity College of the University of Cambridge and went to receive his Ph.D. in biochemistry and pharmacology in 1964. The following year he came to the U.S. on a postdoctoral fellowship in the laboratories of Julius Axelrod at the National Institute of Mental Health and of Steven Kuffler at the Harvard Medical School. He then returned to England to be-

come a research fellow in the department of pharmacology at Cambridge. In 1967 he was elected a Locke Research Fellow of the Royal Society, and in 1971 he became director of the Neurochemical Pharmacology Unit. Iversen is chief editor of the *Journal of Neurochemistry* and president of the European Neuroscience Association.

FLOYD E. BLOOM ("Neuropeptides") is director and member of the division of preclinical neuroscience and endocrinology of Scripps Clinic and Research Foundation. He is a graduate of Southern Methodist University and the Washington University School of Medicine. Before being appointed to his current position in 1975 he was chief of the laboratory of neuropharmacology and acting director of the Division of Special Mental Health Research Programs of the National Institute of Mental Health at St. Elizabeth's Hospital in Washington. His interest in brain research, he says, "stemmed directly from my medical interest in how the brain monitors and controls blood pressure. The continued pursuit of that simple question has led me to develop and apply a variety of experimental approaches to the task of identifying where and how chemical transmitters communicate messages between nerve cells and how those signals are able to alter behavior."

RICHARD J. WURTMAN is a professor of neuroendocrine regulation at the Massachusetts Institute of Technology. His bachelor's degree was granted by the University of Pennsylvania in 1956 and his M.D. by the Harvard Medical School in 1960. He was a medical research officer at the National Institute of Mental Health from 1965 to 1967. In the latter year he moved to M.I.T. In addition to the subject of the current article Wurtman's research interests include the catecholamines, acetylcholine, glutamate and the hormones that cause menarche and ovulation.

W. MAXWELL COWAN ("The Development of the Brain") is vice president of scientific affairs and director of the Developmental Neurobiology Laboratory at Salk

Institute. A native of South Africa, he received his undergraduate education at the University of the Witwatersrand. In 1953 he went to the University of Oxford to obtain his Ph.D. and complete his medical training, and from 1958 until 1966 he was a fellow of Pembroke College. In 1966 he emigrated to the U.S. and taught at the University of Wisconsin; after two years he moved to Washington University. Much of Cowan's scientific work has been on the organization of the limbic system and the development of the visual system.

ROBERT H. WURTZ, MICHAEL E. GOLDBERG and DAVID LEE ROBINSON are colleagues at the Laboratory of Sensorimotor Research at the National Eye Institute who have a common interest in visual perception. Wurtz is chief of the laboratory. He received his bachelor's degree in 1958 at Oberlin College and his doctorate in 1962 from the University of Michigan. From 1962 to 1965 he was research associate at Washington University. Wurtz moved to the National Institutes of Health (NIH) in 1965 and was appointed to his present position in 1978. Goldberg is chief of the section on basic neuro-ophthalmologic mechanisms at the laboratory. His A.B. (1963) is from Harvard College; his M.D. (1968) is from the Harvard Medical School. In 1969 he joined the staff of the NIH as associate neurologist at the National Institute of Mental Health. He left to serve a residency and to work at the Armed Forces Radiobiological Research Institute, returning to the NIH in 1978. Since 1976 he has also been clinical associate professor of neurology at Georgetown University. Robinson is research physiologist at the laboratory. He obtained his B.S. (1965) at Springfield College, his M.S. (1968) at Wake Forest University and his Ph.D. (1972) from the University of Rochester. From 1971 to 1974 he was research fellow in neurophysiology at the National Institute of Mental Health, leaving in 1974 to become research physiologist at the Armed Forces Radiobiological Research Institute. He returned to the NIH in 1978.

TOMASO POGGIO ("Vision by Man and Machine") is professor in the department of psychology and the Artificial Intelligence Laboratory and director of the Center for Biological Information Processing at the Massachusetts Institute of Technology. He writes: "After completing my Ph.D. in theoretical physics at the Max Planck Institute for Biological Cybernetics in Tübingen to work with Werner Reichardt on the amazingly smart and small visual system of the fly. In 1976 I began collaborating with the late David Marr of M.I.T. on the computational approach to vision. At the same time I undertook a complementary kind of work: trying to understand the information-processing mechanism of the brain and its basis in the biophysical properties of nerve cells. In 1981 I moved to M.I.T. to continue work

on computational problems in vision and their applications to robotics and biological information processing."

ERIC I. KNUDSEN ("The Hearing of the Barn Owl") is assistant professor of neurobiology at the Stanford University School of Medicine. He earned his degrees in the University of California system: an A.B. in 1971 and an M.A. in 1973 (both at Santa Barbara) and a Ph.D. in 1976 (from San Diego). He began his research career, he reports, "as an undergraduate working on the bioluminescence of the sea pansy, a primitive animal in one of the lowest phyla (Coelenterata). Since then my interests have moved progressively up the phylogenetic ladder, from the horseshoe crab (master's thesis) to the catfish (doctoral thesis) to the owl (postdoctoral work). I fully expect to be studying mammals, including human beings, before long."

RICHARD H. SCHELLER and RICHARD AXEL ("How Genes Control an Innate Behavior") are assistant professor of biological sciences at Stanford University and professor of pathology and biochemistry at the Columbia University College of Physicians and Surgeons. After Scheller's graduation from the University of Wisconsin at Madison, he went to the California Institute of Technology, where he got his Ph.D. in 1980. In 1980 and 1981 he was a postdoctoral fellow in the laboratories of Axel and Eric Kandel at the College of Physicians and Surgeons. Axel received his A.B. from Columbia College and his M.D. from the Johns Hopkins School of Medicine. He joined the Columbia faculty in 1972. The main theme of his work has been the development of universal techniques for transferring genes from virtually any cell to any other cell.

DANIEL C. TOSTESON ("Lithium and Mania") is dean of the Harvard Medical School, where he is also Caroline Shields Walker Professor of Physiology. He was graduated from Harvard College in 1944 and from the Harvard Medical School in 1949. After completing his internship and residency at Presbyterian Hospital in New York, he worked as a research fellow at the Brookhaven National Laboratory, the National Heart Institute, the Biological Isotope Research Laboratory in Copenhagen and the Physiological Laboratory in Cambridge, England. He began his academic career in 1958 as associate professor of physiology at the Washington University School of Medicine and moved on in 1961 to become professor and chairman of the department of physiology at the Duke University Medical Center. Before returning to Harvard in 1977 he was Lowell T. Coggeshall Professor of Medical Sciences and dean of the Pritzker School of Medicine and vice-president of the Medical Center of the University of Chicago. His research, Tosteson writes, "is directed toward understanding the cellular functions and molecular mechanisms of ion transport across membranes."

JERRY A. FODOR is chairman of the philosophy department of the Massachusetts Institute of Technology. He got his B.A. at Columbia College and his Ph.D. in philosophy from Princeton University. He writes: "I have been a member of the faculty at M.I.T. more or less continuously since 1960, and I now hold a joint appointment as professor in the department of psychology and in the (recently formed) department of linguistics and philosophy. I have written books and articles in all three fields. My current interests include (in approximately reverse order) problems in the theory of meaning, experimental issues in psycholinguistics and cognitive psychology, problems in the philosophy of mind, and sailboats."

BIBLIOGRAPHIES

I NEURON AND SYNAPSE

1. The Neuron

FROM NEURON TO BRAIN: A CELLULAR APPROACH TO THE FUNCTION OF THE NERVOUS SYSTEM. Stephen W. Kuffler and John G. Nicholls. Sinauer Associates, Inc., Publishers, 1976.

NEUROMUSCULAR TRANSMISSION. J. H. Steinback and C. F. Stevens in *Frog Neurobiology: A Handbook,* edited by R. Llinás and W. Precht. Springer-Verlag, 1976.

GATING IN SODIUM CHANNELS OF NERVE. Bertil Hilli in *Annual Review of Physiology,* Vol. 38, pages 139–152; 1978.

INTERACTIONS BETWEEN INTRINSIC MEMBRANE PROTEIN AND ELECTRIC FIELD: AN APPROACH TO STUDYING NERVE EXCITABILITY. Charles F. Stevens in *Biophysical Journal,* Vol. 22, No. 2, pages 295–306; May, 1978.

CONTROL OF ACETYLCHOLINE RECEPTORS IN SKELETAL MUSCLE. Douglas M. Fambrough in *Physiological Reviews,* Vol. 59, No. 1, pages 165–227; January, 1979.

2. Calcium in Synaptic Transmission

A STUDY OF SYNAPTIC TRANSMISSION IN THE ABSENCE OF NERVE IMPULSES. B. Katz and R. Miledi in *Journal of Physiology,* Vol. 192, No. 2, pages 407–436; September, 1967.

MEMBRANE ULTRASTRUCTURE OF THE GIANT SYNAPSE OF THE SQUID *LOLIOGO PEALII*. D W. Pumplin and T. S. Reese in *Nueroscience,* Vol. 3, No. 8, pages 685–696; August, 1978.

PRESYNAPTIC CALCIUM CURRENTS IN SQUID GIANT SYNAPSE. R. Llinás, I. Z. Steinberg and K. Walton in *Biophysical Journal,* Vol. 33, pages 289–321; March, 1981.

RELATIONSHIP BETWEEN PRESYNAPTIC CALCIUM CURRENT AND POSTSYNAPTIC POTENTIAL IN SQUID GIANT SYNAPSE. R. Llinás, I. Z. Steinberg and K. Walton in *Biophysical Journal,* Vol. 33, pages 323–351; March, 1981.

TRANSMISSION BY PRESYNAPTIC SPIKELIKE DEPOLARIZATION IN THE SQUID GIANT SYNAPSE. R. Llinás, M. Suigmori and S. N. Simon in *Proceedings of the National Academy of Sciences of the United States of America,* Vol. 79, No. 7, pages 2415–2419; April, 1982.

II CHEMISTRY OF THE BRAIN

3. The Chemistry of the Brain

CHEMISTRY OF SYNAPTIC TRANSMISSION: ESSAYS AND SOURCES. Edited by Zach W. Hall, John G. Hildebrand and Edward A. Kravitz. Chiron Press, 1974.

HANDBOOK OF PSYCHOPHARMACOLOGY. Edited by Leslie L. Iversen, Susan D. Iversen and Solomon H. Snyder. Plenum Press, 1975–78.

THE BIOCHEMICAL BASIS OF NEUROPHARMACOLOGY. Jack R. Cooper, Floyd E. Bloom and Robert H. Roth. Oxford University Press, 1978.

CENTRALLY ACTING PEPTIDES. Edited by J. Hughes. University Park Press, 1978.

BEHAVIORAL PHARMACOLOGY. Susan D. Iversen and Leslie L. Iversen. Oxford University Press, second edition in press.

4. Neuropeptides

PEPTIDES: INTERGRATORS OF CELL AND TISSUE FUNCTION. Edited by Floyd E. Bloom. Raven Press, 1980.

PEPTIDERGIC NEURONS. Tomas Hökfelt, Olle Johansson, Åke Ljungdahl, Jan M. Lundberg and Marianne Schulzberg in *Nature*, Vol. 284, No. 5756, pages 515–521; April 10, 1980.

BRAIN PEPTIDES AS NEUROTRANSMITTERS. Solomon H. Snyder in *Science*, Vol. 209, No. 4460, pages 976–983; August 29, 1980.

5. Nutrients That Modify Brain Function

NUTRITION AND THE BRAIN. Edited by Richard J. Wurtman and Judith J. Wurtman. Raven Press, 1977–82.

RELEASE OF ACETYLCHOLINE FROM THE VASCULAR PERFUSED RAT PHRENIC NERVE-HEMIDIAPHRAGM. George C. Bierkamper and Alan M. Goldberg in *Brain Research*, Vol. 202, No. 1, pages 234–237; November 24, 1980.

PRECURSOR CONTROL OF NEUROTRANSMITTER SYNTHESIS. R. J. Wurtman, F. Hefti and E. Melamed in *Pharmacological Reviews*, Vol. 32, No. 4, pages 315–335; December, 1980.

ALZHEIMER'S DISEASE: A REPORT OF PROGRESS IN RESEARCH. Edited by S. Corkin, K. J. Davis, J. H. Growdon, E. Usdin and R. J. Wurtman. Raven Press, 1981.

CARBOHYDRATE CRAVING IN OBESE PEOPLE: SUPPRESSION BY TREATMENTS AFFECTING SEROTONINERGIC TRANSMISSION. Judith J. Wurtman, Richard J. Wurtman, John H. Growdon, Peter Henry, Anne Lipscomb and Steven H. Zeisel in *International Journal of Eating Disorders*, Vol. 1, No. 1; pages 2–15; Autumn, 1981.

III DEVELOPMENT OF THE BRAIN

6. The Development of the Brain

DEVELOPMENTAL PROGRAMMING FOR RETINOTECTAL PATTERNS. R. Kevin Hunt in *Cell Patterning: Ciba Foundation Symposium 29*. Associated Scientific Publishers, 1975.

CELL MIGRATION AND NEURONAL ECTOPIAS IN THE BRAIN. P. Rakic in *Birth Defects: Original Articles Series*, Vol. 11, pages 95–129; 1975.

PLASTICITY OF OCULAR DOMINANCE COLUMNS IN MONKEY STRIATE CORTEX. D. H. Hubel, T. N. Wiesel and S. LeVay in *Philosophical Transactions of the Royal Society of London, Series B*, Vol. 278, No. 961, pages 377–409; April 26, 1977.

IV SENSORY PROCESSES

7. Brain Mechanisms of Visual Attention

BEHAVIORAL MODULATION OF VISUAL RESPONSES IN THE MONKEY: STIMULUS SELECTION FOR ATTENTION AND MOVEMENT. Robert H. Wurtz, Michael E. Goldberg and David Lee Robinson in *Progress in Psychobiology and Physiological Psychology*, Vol. 9, pages 43–83; 1980.

BEHAVIORAL ENHANCEMENT OF VISUAL RESPONSES IN MONKEY CEREBRAL CORTEX, I: MODULATION IN POSTERIOR PARIETAL CORTEX RELATED TO SELECTIVE VISUAL ATTENTION. M. Catherine Bushnell, Michale E. Goldberg and David Lee Robinson in *Journal of Neurophysiology*, Vol. 46, No. 4, pages 755–772; October, 1981.

8. Vision by Man and Machine

ARTIFICIAL INTELLIGENCE. Patrick Henry Winston. Addison-Wesley Publishing Co., 1977.

A COMPUTATIONAL THEORY OF HUMAN STEREO VISION. D. Marr and T. Poggio in *Proceedings of the Royal Society of London, B*, Vol. 204, No. 1156, pages 301–328; May 23, 1979.

FROM IMAGES TO SURFACES: A COMPUTATIONAL STUDY OF THE HUMAN EARLY VISUAL SYSTEM. William Eric Leifur Grimson. The MIT Press, 1981.

THEORETICAL APPROACHES IN NEUROBIOLOGY. Edited by Werner E. Reichardt and Tomaso Poggio. The MIT Press, 1981.

INTENSITY, VISIBLE-SURFACE, AND VOLUMETRIC REPRESENTATIONS. H. K. Nishihara in *Artificial Intelligence*, Vol. 17, Nos. 1–3, pages 265–284; August, 1981.

VISION, David Marr. W. H. Freeman and Company, 1982.

9. The Hearing of the Barn Owl

ACOUSTIC LOCATION OF PREY BY BARN OWLS (*TYTO ALBA*). Roger S. Payne in *The Journal of Experimental Biology*, Vol. 54, pages 535–573; 1971.

HOW THE OWL TRACKS ITS PREY. Masakazu Konishi in *American Scientist*, Vol. 61, No. 4, pages 414–424; July–August, 1973.

SOUND LOCATION IN BIRDS. E. I. Knudsen in *Comparative Studies of Hearing in Vertebrates*, edited by Arthur N. Popper and Richard R. Fay. Springer-Verlag, 1980.

V   BRAIN, BEHAVIOR, AND MIND

10. How Genes Control an Innate Behavior

A FAMILY OF GENES THAT CODES FOR ELH, A NEUROPEPTIDE ELICITING A STEREOTYPED PATTERN OF BEHAVIOR IN APLYSIA. Richard H. Scheller, James F. Jackson, Linda Beth McAllister, James H. Schwartz, Eric R. Kandel and Richard Axel in *Cell*, Vol. 28, No. 4, pages 707–719; April, 1982.

A SINGLE GENE ENCODES MULTIPLE NEUROPEPTIDES MEDIATING A STEREOTYPED BEHAVIOR. Richard H. Scheller, James F. Jackson, Linda B. McAllister, Barry S. Rothman, Earl Mayeri and Richard Axel in *Cell*, Vol. 32, No. 1, pages 7–22; January, 1983.

IN SITU HYBRIDIZATION TO STUDY THE ORIGIN AND FATE OF IDENTIFIED NEURONS. Linda B. McAllister, Richard H. Scheller, Eric R. Kandel and Richard Axel in *Science*, Vol. 222, No. 4625, pages 800–808; November 18, 1983.

11. Lithium and Mania

LITHIUM SALTS IN THE TREATMENT OF PSYCHOTIC EXCITEMENT. J. F. J. Cade in *The Medical Journal of Australia*. Vol. 36, pages 349–352; 1949.

LITHIUM IN MEDICAL PRACTICE. Edited by F. N. Johnson. University Park Press, 1978.

KINETICS AND STOICHIOMETRY OF NA-DEPENDENT LI TRANSPORT IN HUMAN RED BLOOD CELLS. B. Sarkadi, J. K. Alifimoff, R. B. Gunn and D. C. Tosteson in *Journal of General Physiology*, Vol. 72, No. 2, pages 249–265; August, 1978.

STUDIES OF LITHIUM TRANSPORT ACROSS THE RED CELL MEMBRANE, V: ON THE NATURE OF THE Na^+-DEPENDENT Li^+ COUNTERTRANSPORT SYSTEM OF MAMMALIAN ERYTHROCYTES. Jochen Duhm and Bernhard F. Bechker in *Journal of Membrane Biology*, Vol. 51, No. 3/4, pages 263–286; 1979.

LITHIUM TRANSPORT IN HUMAN RED BLOOD CELLS: GENETIC AND CLINICAL ASPECTS. Ghanshyam N. Pandey, Elizabeth Dorus, John M. Davis and Daniel C. Tosteson in *Archives of General Psychiatry*, Vol. 36, No. 8, pages 902–908; July 20, 1979.

LITHIUM, MEMBRANES, AND MANIC-DEPPRESIVE ILLNESS. Barbara E. Ehrlich and Jared M. Diamond in *The Journal of Membrane Biology*, Vol. 52, No. 3, pages 187–200; 1980.

12. The Mind-Body Problem

PSYCHOLOGICAL EXPLANATION: AN INTRODUCTION TO THE PHILOSOPHY OF PSYCHOLOGY. Jerry A. Fodor. Random House, Inc., 1968.

RES COGITANS: AN ESSAY IN RATIONAL PSYCHOLOGY. Zeno Vendler. Cornell University Press, 1972.

THE LANGUAGE OF THOUGHT. Jerry A. Fodor. Thomas Y. Crowell Co., 1975.

MIND, LANGUAGE AND REALITY: PHILOSOPHICAL PAPERS, VOL. 2. Hilary Putnam. Cambridge University Press, 1975.

READINGS IN PHILOSOPHY OF PSYCHOLOGY: VOL. 1. Edited by N. Block. Harvard University Press, 1980.

INDEX